RENEWABLE ENERGY TARIFFS AND INCENTIVES IN INDONESIA

REVIEW AND RECOMMENDATIONS

SEPTEMBER 2020

ASIAN DEVELOPMENT BANK

ADB

© 2020 Asian Development Bank
6 ADB Avenue, Mandaluyong City, 1550 Metro Manila, Philippines
Tel +63 2 8632 4444; Fax +63 2 8636 2444
www.adb.org

Some rights reserved. Published in 2020.

ISBN 978-92-9262-323-4 (print); 978-92-9262-324-1 (electronic); 978-92-9262-325-8 (ebook)
Publication Stock No. TCS200254
DOI: http://dx.doi.org/10.22617/TCS200254

The views expressed in this publication are those of the authors and do not necessarily reflect the views and policies of the Asian Development Bank (ADB) or its Board of Governors or the governments they represent.

ADB does not guarantee the accuracy of the data included in this publication and accepts no responsibility for any consequence of their use. The mention of specific companies or products of manufacturers does not imply that they are endorsed or recommended by ADB in preference to others of a similar nature that are not mentioned.

By making any designation of or reference to a particular territory or geographic area, or by using the term "country" in this document, ADB does not intend to make any judgments as to the legal or other status of any territory or area.

Corrigenda to ADB publications may be found at http://www.adb.org/publications/corrigenda.

Note:
In this publication, "$" refers to United States dollars.

Cover design by Editha Creus.

This report was prepared by PT. Castlerock and Economic Consulting Associates for the Ministry of Finance of Indonesia.

Contents

Tables and Figures

Abbreviations

Names for Indonesian entities and legal documents are given in both English and Indonesian on first use. Subsequent references use the Indonesian acronym. An exception is made for the Ministry of Finance where the English acronym is used given its familiarity.

ADB	Asian Development Bank
APBN	Anggaran Pendapatan dan Belanja Negara (State Budget)
BOOT	build–own–operate–transfer
BPP	biaya pokok produksi (production cost)
CCGT	combined-cycle gas turbine
BPDLH	Badan Pengelola Dana Lingkungan Hidup (Environmental Fund Management Agency)
DPR	Dewan Perwakilan Rakyat (House of Representatives)
ESDM	Energi dan Sumber Daya Mineral (Ministry of Energy and Mineral Resources)
FIT	feed-in tariff
KEN	Kebijakan Energi Nasional (National Energy Policy)
IPP	Independent power producer
KEPMEN	Keputusan Menteri (Ministerial Decree)
MOF	Ministry of Finance
PERMEN	Peraturan Menteri (Ministerial Regulation)
PERPRES	Peraturan Presiden (Presidential Regulation)
PLN	Perusahaan Listrik Negara (State Electricity Company)
PMK	Peraturan Menteri Keuangan (Minister of Finance Regulation)
PPA	power purchase agreement
PPP	public–private partnership
PSO	public service obligation
PT SMI	PT Sarana Multi Infrastruktur
PV	photovoltaic
RE	renewable energy
RUEN	Rencana Umum Energi Nasional (National Energy Plan)
RUPTL	Rencana Usaha Penyediaan Tenaga Listrik (Electricity Supply Business Plan)
SB/SO	Single buyer/system operator
SOE	state-owned enterprise

Weights and Measures

GW	gigawatt
GWh	gigawatt-hour
kW	kilowatt
kWh	kilowatt-hour
MW	megawatt
tCO_2e	ton of carbon dioxide equivalent
TWh	terawatt-hour

Currency Equivalents

(as of 7 July 2020)

Currency unit	–	rupiah (Rp)
$1.00	=	Rp14,419.80
Rp1.00	=	$0.000069

Acknowledgments

This work was conducted under the technical supervision of Florian Kitt, energy specialist, Energy Division, Southeast Asia Department (SERD), of the Asian Development Bank (ADB). Additional guidance and support was provided by Andrew Jeffries (director, Energy Division, SERD, ADB). The report was written by William Derbyshire (consultant, ADB) and Mike Crosetti (consultant, ADB), with contributions from Kelsey Yates (consultant, ADB). The team wishes to thank Yongping Zhai (chief, Energy Sector Group, Sustainable Development and Climate Change Department, ADB) as the report's peer reviewer.

The team wishes to thank the Indonesian Ministry of Finance, PT Sarana Multi Infrastruktur, and development partners for their inputs and discussions during the preparation of the report. The team also thanks colleagues from the Indonesia Resident Mission and the Department of Communications for their support.

Director	Andrew Jeffries, Energy Division, SERD
Team leader	Florian Kitt, energy specialist, SERD
Team member	Jeffrey Almera, senior operations assistant, SERD
Peer reviewer	Yongping Zhai, chief, Energy Sector Group, Sustainable Development and Climate Change Department

Key Recommendations

This report proposes a renewable energy (RE) subsidy mechanism to close the gap between the costs of renewable power and conventional power generation, taking into account the additional economic benefits of renewable power for Indonesia. The subsidy should be calculated as the difference between the cost of supply from a given renewable power project and the financial cost that the Perusahaan Listrik Negara (PLN, State Electricity Company) would have otherwise incurred for generation on that system in the absence of the renewable project, i.e., PLN's "avoided cost." To ensure that the Government of Indonesia does not overpay for renewable subsidies, the cost of renewable supply would be capped at its economic value, which is calculated as the economic avoided cost plus the social benefits of externalities.

If Indonesia can adopt international best practice for RE planning, procurement, contracting, and risk mitigation, then the financial costs of RE development will decline accordingly. Until now the government has not adequately taken into account the dependency of RE costs on the broader regulatory and commercial environment. The implementation of this subsidy should therefore be part of a broader interministerial electricity policy reform program.

Mechanism Component	Recommended Approach
Legal basis	• Compensate PLN for a public service obligation to procure renewable generation to meet national targets.
Governance	• PLN prepares annual subsidy estimate for review and approval by the Energi dan Sumber Daya Mineral (ESDM, Ministry of Energy and Mineral Resources), the Ministry of Finance, and ultimately the Dewan Perwakilan Rakyat (DPR, House of Representatives) for budgeting in the Anggaran Pendapatan dan Belanja Negara (APBN, State Budget). • "True-up" at the end of the year for differences between forecast and actual (audited) RE project prices and volumes.
Subsidy calculation	• Calculate individually for each RE project and then sum to obtain the total annual RE subsidy estimate. • The subsidy is the difference between the RE project price and PLN's financial cost savings from the RE project (as estimated at the date of power purchase agreement [PPA] signature) multiplied by PLN's actual purchases (or production in the case of PLN's own plant). • A margin is added to the RE subsidy, consistent with legal requirements and current practice with the tariff subsidy. A substantially lower margin than the prevailing 7% margin applied to the tariff subsidy in the order of 0.5% is recommended, consistent with the administrative and financial costs PLN actually incurs.

Mechanism Component	Recommended Approach

Subsidy calculation *(continued)*

Cap on PPA Price

- If the PPA price results from a competitive tender, then the RE project cost used for subsidy calculations is capped at the economic value.
- If there is no competitive tender, then the allowed RE project price is capped at the lower threshold of the estimated production cost or economic value, calculated separately for each RE technology and PLN grid. This price serves as the PPA price and is not subject to further negotiation by PLN or government.
- The government can also apply a cap on total subsidies payable to maintain fiscal prudence. RE projects would be developed up to this cap.

Implementation issues	• Amend existing ESDM regulations to allow PLN to sign PPAs at prices above the *biaya pokok produksi* (BPP, production cost) –based cap, if a subsidy is available. • The existing ESDM regulations will also need to be amended to replace the use of BPP with estimated financial cost savings and, ideally, to allow the use of auctions as a preferred procurement mechanism.

Executive Summary

Background and Context

Despite the availability of financing and the existence of tax incentives for renewable power development, Indonesia is failing to meet its renewable energy (RE) targets. There are multiple causes of this failure, and other impediments will also need to be addressed. Among these are (i) inadequate power system planning and grid management practices; (ii) unbalanced power purchase agreements that adversely affect bankability; (iii) counterproductive procurement and contracting processes, negotiation practices, build–own–operate–transfer (BOOT) requirements, and change-of-ownership restrictions; (iv) high local content requirements prior to establishment of a market large enough to achieve domestic manufacturing economies of scale; and (v) limitations on foreign investment.

The relative importance of these impediments depends on the particular renewable generation technology, but among the most damaging are the disincentives or outright prohibitions against the State Electricity Company (PLN) purchasing renewable power at prices higher than conventional alternatives.

The Ministry of Finance (MOF) is particularly interested in accelerating geothermal power development as it is a predominant source of renewable energy in Indonesia, representing 44% of the nation's actual renewable power production in 2018 and 42% of PLN's 2028 renewable power generation forecast. It is the focus of this report.

Current regulations cap the price paid for geothermal purchases outside of Java and Sumatera at PLN's generation production cost (BPP), or average accounting cost of production over the past year on those systems. However, most of Indonesia's geothermal potential is found on the islands of Java and Sumatera. In principle, prevailing regulation allows PLN to freely negotiate the price of geothermal power with developers on these islands. In reality, however, the caps on PLN's own retail tariffs provide a strong disincentive for PLN to purchase anything but the lowest-cost source of power. Under current conditions on PLN's large power systems, coal-fired generation typically has far lower financial costs than geothermal, and is therefore PLN's preferred candidate.

This report proposes an RE subsidy mechanism to close the gap between the costs of renewable power and conventional power generation, taking into account the additional economic benefits of renewable power for Indonesia.

Closing the Price–Cost Gap

The Government of Indonesia has not adjusted PLN's retail tariffs since 2017 to help ensure the affordability of electricity. The government also capped the price of renewables to help preserve PLN's financial viability under these fixed tariffs. As tariff affordability and the financial health of PLN will continue to be major concerns of the government, there is a need to close the gap between prices and costs through a renewable energy subsidy to PLN. This would be consistent with both Law 19/2003 on State-Owned Enterprises and Law 30/2007 on Energy.

The subsidy should be calculated as the difference between the cost of supply from a given renewable power project and the financial cost that PLN would have otherwise incurred for generation on that system in the absence of the renewable project, i.e. PLN's "avoided cost." To ensure that the government does not overpay for renewable subsidies, the cost of renewable supply would be capped at its economic value, which is calculated as the economic avoided cost plus the social benefits of externalities.

This subsidy framework requires determination of three values:

i. **PLN's financial avoided cost.** Contrary to the current approach which utilizes PLN's region-specific average historical accounting cost of generation (BPP) as a cost benchmark, PLN's financial avoided cost should be determined on the basis of a forward-looking marginal cost corresponding to the most likely plant to be added in the future. This will vary by system because the most likely conventional plant addition depends on system size, load characteristics, available resources, etc. For example, for large power systems like Java-Bali, generation options include ultra-super critical coal-fired or large gas-fired combined cycle plants, whereas on small remote island system diesel generation might be used. This represents PLN's cost savings from renewables purchases.

ii. **The financial cost of supply from the renewable project,** which is the price required for the RE project to recover its costs, including a reasonable return.

    a. This should be determined on a competitive basis through reverse auctions or similar methods. The direct selection method currently applied by PLN does not provide sufficient competition to be used for this purpose.

    b. Where reverse auctions are not possible or appropriate (e.g., geothermal projects, for which areas of work are awarded based on the best exploration offer, or site-specific hydropower projects) then the government should apply a cost estimation methodology that calculates the price based on key project parameters like resource size and quality. This would provide a "sliding scale" that determines the cost based on key project cost drivers outside the control of the developer.

iii. The economic value ceiling of renewable supply, which is calculated as the sum of the economic avoided cost to PLN plus the value of externalities such as domestic health impacts, global environmental impacts, energy security considerations, local employment impacts, etc. This represents the maximum value that Indonesia should pay for renewables purchases.

The subsidy is therefore calculated on a project-by-project basis based on actual RE project output as follows:

(RE project financial cost/kWh – PLN avoided cost/kWh) x (RE project kWh output)

provided that: RE project financial cost/kWh < economic value/kWh

Regardless of whether a reverse auction or the government's production cost estimate is used to set the RE project financial cost, the resulting value should be applied without further negotiation, and the subsidy paid to PLN on that basis. This would greatly accelerate RE development by removing the need for PLN and/or the Ministry of Energy and Mineral Resources (ESDM) to individually negotiate prices for each project, which has proven to be a very slow process. Moreover, in the case of reverse auctions, if developers anticipate that their bids will be subject to subsequent negotiation, they will price that into their bids.

The record low prices for RE achieved around the world through reverse auctions without subsequent negotiation demonstrates the effectiveness of this approach. These results also rely on a well-structured tender process, well-balanced power purchase agreements (PPAs) and other measures that would need to accompany the introduction of the RE subsidy in order to achieve the lowest prices.

When reverse auctions cannot be used, there may be concerns that government-stipulated prices could result in payments to project sponsors higher than the minimum price they would accept, i.e., that government might overpay. The use of a production cost model rather than a feed-in tariff (FIT) minimizes this risk for technologies like geothermal that are characterized by numerous project-specific cost drivers outside of a developer's control. A production cost model provides far greater granularity in terms of setting the project price because it can take into account many more variables than an FIT regime. Moreover, it can take into account the specific value of a given variable for a given project, rather than having to determine a project price based on the broad ranges of values typically used for FIT schemes.

Subsidy Implementation

PLN will propose annually to ESDM the required subsidy based on expected RE additions in the coming year plus operation of existing projects developed previously under this framework, as well as financial avoided costs for each system and estimated financial costs of supply for each new RE project.

ESDM and MOF will review these and ensure that none of the RE financial costs of supply exceed the economic value ceiling. For upcoming RE projects that will not be procured through competitive tender, MOF will estimate the RE project production cost with its own technology-specific cost model that takes into account specific project characteristics, e.g., resource size and quality. There should be a single production cost model used across government. Once approved by ESDM, MOF, and the House of Representatives (DPR), these subsidies will be budgeted in the State Budget (APBN).

As with the existing electricity tariff subsidy, the renewable subsidy will be paid in arrears with a final "true-up" based on actual generation output from the eligible RE projects. As per Law 19/2003 on State Enterprises, a margin will also be applied to the difference between the RE project costs and PLN avoided costs. However, this margin would be intended to cover only PLN's administration and time value of money (arising from the mismatch between power purchases and subsidy receipts), so it is expected to be at a much lower level than the margin applied for the tariff subsidy (currently 7%). MOF and ESDM should annually verify the integrity and reasonableness of the RE project cost estimate inputs and PLN avoided costs through annual external review with opportunities for stakeholder review and input.

It is understood that PERMEN (Ministerial Regulation) ESDM 50/2017 will be replaced. Any new regulation should allow PLN to pass on government subsidies for RE projects to project sponsors, or to receive RE subsidies for its own RE projects.

Estimated Subsidy Payments

Two scenarios were prepared to estimate future subsidy payments to new geothermal projects under the recommended scheme. The first scenario utilized the geothermal capacity additions planned in PLN's Rencana Usaha Penyediaan Tenaga Listrik (RUPTL, Electricity Supply Business Plan) 2019–2028. Given that over the past 15 years PLN has never achieved its geothermal capacity expansion plan, a second scenario was prepared which excluded all geothermal capacity additions in the business plan for which prospects had not been specifically identified. In addition, the capacity for named prospects that have not yet been proven was adjusted to the estimates presented in an earlier study of geothermal potential in Indonesia conducted by Castlerock Consulting, prepared for the Ministry of Energy and Mineral Resources (ESDM) and included a detailed assessment of 50 geothermal prospects across the country. Both scenarios considered only capacity additions for Java and Sumatera, as these systems represent 83% of all geothermal capacity additions planned in the RUPTL 2019–2028, and are systems where geothermal will most likely require subsidies.

For both scenarios, a detailed geothermal production cost model was applied to estimate the financial cost of supply from each project. This model was prepared by Jim Randle and Jim Lawless for the Directorate General of New and Renewable Energy and Energy Conservations under funding from the New Zealand Ministry of Foreign Affairs and Trade, and subsequently updated by the authors for use by PT Sarana Multi Infrastruktur and others. PLN's financial avoided costs were estimated based on plants identified in the business plan, and economic cost ceilings established based on analysis from other sources.

The RUPTL scenario results in an annual geothermal subsidy payment of Rp17.9 trillion for 3,786 megawatts (MW) of new capacity in 2028. In contrast, by 2028, the adjusted scenario results in an annual subsidy payment of Rp6.3 trillion for 1,546 MW of new capacity.

The government wishes to minimize subsidies and budget future subsidy obligations with certainty. One way of reducing subsidies is to simply apply a cap on additional subsidies to be awarded in a year or over a number of years and allocate this through reverse auctions. This will reduce the amount of capacity brought online but will allow the lowest-cost projects to proceed. Where auctions are not possible, other mechanisms might be adopted, such as first-come-first-served to encourage the most rapidly deployed projects, or ranking by calculated production cost to select the lowest-cost projects.

If the longevity of subsidy obligations is a concern, the government could stage or front-load the subsidy payments, though this would result in higher near-term demands on the state budget rather than spreading payments over the life of the PPA.

But the most effective way to reduce potential subsidy obligations, and more generally support the effectiveness of this subsidy mechanism to accelerate RE development (particularly for technologies other than geothermal that are already becoming competitive with fossil fuels in terms of cost), is for the government to address all of the other impediments mentioned earlier.

1. Introduction and Background

If Indonesia can move to international best practice for renewable energy planning, procurement, contracting, and risk mitigation, then the financial costs of RE development will decline accordingly. Until now, the government has not adequately taken into account the dependency of RE costs on the broader regulatory and commercial environment. The implementation of this subsidy should therefore be part of a broader interministerial electricity policy reform program.

1.1 Purpose

The Ministry of Finance (MOF) wishes to assess existing support schemes for renewable energy (RE) development in Indonesia and develop a framework that will accelerate development in line with government targets. This should provide MOF with a clear policy framework on how to best support RE in terms of financing and fiscal mechanisms, as well as a strategy to engage in a discussion on tariff, procurement and contracting policies and practices with line ministries and state-owned enterprises in the electricity sector. MOF is particularly interested in accelerating geothermal power development, as it is a predominant source of renewable energy in Indonesia, representing 44% of the nation's actual renewable power production in 2018 and 42% of PLN's 2028 renewable power generation forecast.

Indonesia has set ambitious targets for expanding electricity generation from RE sources. The Electricity Supply Business Plan (RUPTL) 2019–2028 issued by the State Electricity Company (PLN) aims to increase the share of renewables in installed capacity and generation to 23% by 2025, from 13% in 2017. Meeting these targets requires scaling up current rates of renewables capacity additions by over six times. From 2013 to 2018, total renewables capacity additions averaged 328 megawatts (MW) annually (with 52% being hydro and 33% being geothermal). Achieving the 2025 target will require average annual additions of 1,959 MW (with 44% being hydro and 36% being geothermal).

Given this large gap, this report considers how MOF can support the rapid deployment of renewable energy in the power sector.

1.2 Scope

There are many barriers to the development of RE projects in Indonesia, including foreign investment restrictions, local content requirements and the allocation of risk in power purchase agreements (PPAs). The extent of the existing impediments is readily apparent in the large number of projects which are unable to reach financial close following PPA signature, due to their inability to find investors. Of 75 PPAs signed in 2017–2018, only 35 have reached financial close and moved to construction.[1]

The relative importance of these impediments differs by RE technology. In line with current MOF priorities, this report focuses on barriers to developing geothermal projects. It is important to note that the measures recommended to overcome these impediments may not apply to other RE technologies.

1.3 Structure

This report contains the following sections:

- Section 2 summarizes the current impediments to RE projects, with a focus on pricing arrangements. It also discusses options for closing the price–cost gap, concluding that budget subsidies to PLN are the most feasible solution.

- Section 3 describes how such a budget subsidy mechanism could work as well as the associated challenges created by the existing regulatory arrangements for PLN and RE projects.

- Section 4 sets out next steps, including other desirable reforms, to the regulatory and institutional framework for RE projects.

Further information and analysis supporting the discussion and recommendations are contained in the appendixes.

---

[1]   D. Angriani. 2019. 19 Proyek Listrik Energi Terbarukan Belum Dapat Dana. *Medcom.id.* 15 October. https://www.medcom.id/ekonomi/energi/5b2Ade6N-19-proyek-listrik-energi-terbarukan-belum-dapat-dana.

2. Impediments to Renewable Energy Projects

2.1 Overview

There are many impediments to the development of RE projects in Indonesia: (i) inadequate power system planning and grid management practices; (ii) unbalanced power purchase agreements that adversely affect bankability; (iii) counterproductive procurement and contracting processes, negotiation practices, build–own–operate–transfer (BOOT) requirements, and change-of-ownership restrictions; (iv) high local content requirements prior to establishment of a market large enough to achieve domestic manufacturing economies of scale; and (v) limitations on foreign investment. But among the most damaging are the disincentives or outright prohibitions against PLN purchasing renewable power at prices higher than conventional alternatives. The prevailing renewable energy development environment is described in further detail in Appendix 2.

Most of Indonesia's geothermal potential is found on the islands of Java and Sumatera, and this is where most geothermal development will continue to take place. Prevailing regulation allows PLN to freely negotiate the price of geothermal power with developers on these islands without any constraint. As a practical matter, though, PLN sells most of its power to customers under a strict price cap tariff regime and is thus strongly incentivized to seek the lowest-cost sources of generation. Under current conditions, coal-fired generation typically has lower financial costs than geothermal.

In geothermal resource areas outside of Java and Sumatera, regulation caps the price PLN can pay for geothermal power at PLN's average historical accounting cost of generation on those systems, referred to as average production cost (BPP). Under certain conditions, these price caps may be high enough to facilitate geothermal power development, but small geothermal developments of the kind suitable for smaller grids in these areas can easily exceed these caps.

This report proposes a renewable energy subsidy mechanism to close the gap between the costs of renewable power and conventional power generation, taking into account the additional economic benefits of renewable power for Indonesia. Closing this price–cost gap is necessary but, by itself, not sufficient to deliver the changes needed for Indonesia to realize its renewable energy ambitions. Notably, some RE technologies, as illustrated later in this section, are available at costs similar to or below the current price cap, implying that there are other impediments preventing a more rapid rollout. Key impediments are summarized in Figure 1, before a deeper analysis of the price–cost gap.

Figure 1: Key Impediments to Renewable Energy Development in Indonesia

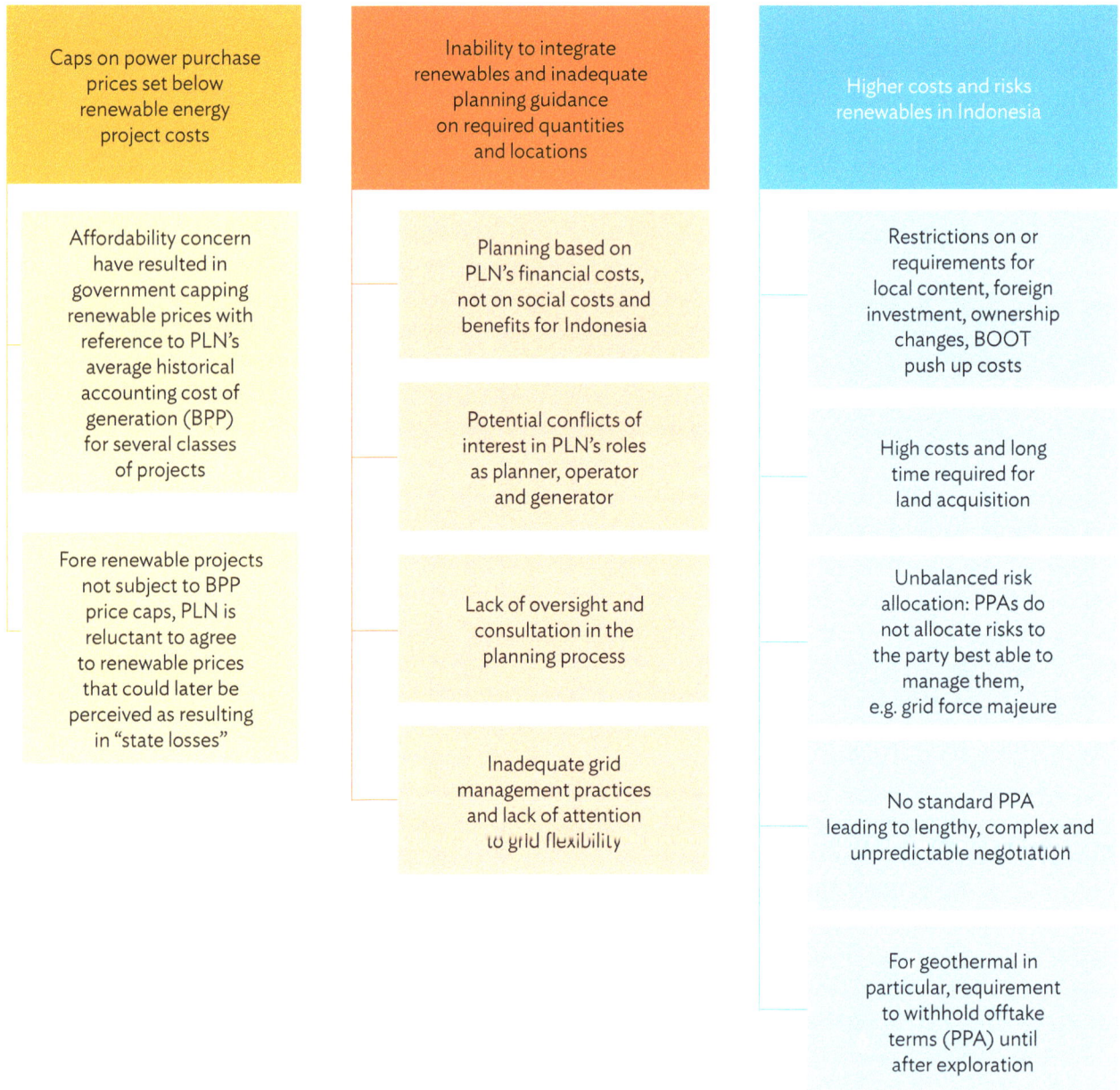

Caps on power purchase prices set below renewable energy project costs	Inability to integrate renewables and inadequate planning guidance on required quantities and locations	Higher costs and risks renewables in Indonesia
Affordability concern have resulted in government capping renewable prices with reference to PLN's average historical accounting cost of generation (BPP) for several classes of projects	Planning based on PLN's financial costs, not on social costs and benefits for Indonesia	Restrictions on or requirements for local content, foreign investment, ownership changes, BOOT push up costs
Fore renewable projects not subject to BPP price caps, PLN is reluctant to agree to renewable prices that could later be perceived as resulting in "state losses"	Potential conflicts of interest in PLN's roles as planner, operator and generator	High costs and long time required for land acquisition
	Lack of oversight and consultation in the planning process	Unbalanced risk allocation: PPAs do not allocate risks to the party best able to manage them, e.g. grid force majeure
	Inadequate grid management practices and lack of attention to grid flexibility	No standard PPA leading to lengthy, complex and unpredictable negotiation
		For geothermal in particular, requirement to withhold offtake terms (PPA) until after exploration

BOOT = build–own–operate–transfer, BPP = *biaya pokok produksi* (production cost),
PLN = Perusahaan Listrik Negara (State Electricity Company), PPA = power purchase agreement,

Source: ADB.

2.2   Price–Cost Gap and Need for Subsidies

2.2.1   Existing Regulations

ESDM is responsible for power sector policy and regulation, which includes regulation of tariffs, as well as procurement and contracting of renewable energy (RE) generation by the State Electricity Company (PLN).

ESDM's overriding power sector policy objective since the start of 2017 has been to ensure affordability of electricity. Given that the Government of Indonesia also aims to reduce subsidies to PLN and simultaneously maintain PLN's financial health, the unwillingness to increase tariffs triggers a need to minimize PLN's costs. As part of this effort, in 2017, ESDM replaced existing regulations on the pricing of RE purchases by PLN with new regulations capping the price of most renewable technologies at some percentage of PLN's generation production cost (BPP). BPP is calculated by PLN region (roughly corresponding to provinces) and small isolated systems based on PLN's accounting cost of generation for the previous year in that region or system.

The key regulation governing RE pricing is **PERMEN ESDM 50/2017 on Utilization of Renewable Energy Resources for Electricity Supply**, as amended by 53/2018.[2] RE power prices are negotiated between PLN and the developer, and subsequently approved by ESDM (which may involve renegotiation). If the project is located in an area where the regional BPP is greater than national average BPP, then the negotiated price cannot exceed either 85% or 100% of the regional BPP depending on the technology. Table 1 summarizes the key provisions regarding pricing, procurement, and form of investment. Where the regional BPP is below the national average BPP then, in principle, no cap applies. In practice, however, PLN is strongly dis-incentivized to contract for any power but the lowest financial cost, which under current conditions on large grids is coal-fired generation.

---

[2]   PERMEN 53/2018 only adds power generation using liquid biofuel to the scope of PERMEN 50/2017.

Table 1:   Current Renewable Energy Pricing (PERMEN 50/2017, amended by 53/2018)

Renewable Energy Resource	Procurement Mechanism	Areas Where Regional BPP < National BPP	Areas Where Regional BPP > National BPP	Form of Investment[3]
Solar	Direct selection with capacity quota	Negotiated between IPP and PLN	Maximum 85% of regional BPP	BOOT
Wind	Direct selection with capacity quota	Negotiated between IPP and PLN	Maximum 85% of regional BPP	BOOT
Hydro	Direct selection	Negotiated between IPP and PLN	Maximum 100% of regional BPP	BOOT
Biomass	Direct selection	Negotiated between IPP and PLN	Maximum 100% of regional BPP	BOOT
Biogas	Direct selection	Negotiated between IPP and PLN	Maximum 100% of regional BPP	BOOT
Waste-to-Energy[a]	Direct appointment by municipality	Negotiated between IPP and PLN	Maximum 100% of regional BPP	Not specified
Geothermal	Direct appointment where resources are proven	Negotiated between IPP and PLN	Maximum 100% of regional BPP	BOOT
Ocean Energy	Direct selection	Negotiated between IPP and PLN	Maximum 100% of regional BPP	BOOT
Biofuel Power Plants	Direct selection	Negotiated between IPP and PLN	Negotiated between IPP and PLN	BOOT

BOOT = build–own–operate–transfer, BPP = *biaya pokok produksi* (production cost), ESDM = *Energi dan Sumber Daya Mineral* (Ministry of Energy and Mineral Resources), IPP = independent power producer, PERMEN = *Peraturan Menteri* (Ministerial Regulation), PERPRES = *Peraturan Presiden* (Presidential Regulation), PLN = *Perusahaan Listrik Negara* (State Electricity Company).

a PERPRES 35/2018 stipulates feed-in tariffs for 12 cities, some of which are higher than BPP.

Source: ADB team summary of PERMEN ESDM 50/2017 and 53/2018.

2.2.2   Issues and Concerns

The current pricing methodology raises three important concerns as regards RE development:

- BPP and prices linked to it are below the cost of some, but not all new RE projects, meaning these projects are not financially viable for developers. Figure 2 compares current BPP and price caps derived from it with estimates of the costs of illustrative new RE projects. Geothermal and wind projects are generally unlikely to be viable at current price caps, but solar and minihydro projects will often be viable at prices below regional BPP, if other impediments to their development can be overcome.

- BPP is based on average historical accounting costs rather than forward-looking marginal costs, which more accurately represent the capital, operating, and fuel costs PLN is likely to face in the future. This makes BPP a poor measure of the cost savings to PLN from RE generation purchases. Specifically:

  - BPP relies on average depreciated book value together with a financing charge allocation to represent capital costs of PLN generation. It is very likely that new fossil fuel generation capacity of the same technology would cost significantly more.

  - There is no margin applied to the BPP calculation. Consequently, BPP does not take into account the equity returns or profit that any developer (including PLN) requires to make new generation investments financially viable.

  - For bulk power purchases, BPP reflects payments to IPPs and PLN subsidiaries. However, PLN subsidiaries are paid at a bulk power purchase tariff significantly lower than a commercially arms-length purchase.

  - For fuel costs, BPP utilizes total fuel costs from the previous year divided by total generation. This changes year-to-year based on fuel markets. It is imprudent to make 20- to 30-year investment decisions based on the previous year's fuel prices rather than a long-term future outlook, all the more given that under the current BPP-based regime the attractiveness of a new project with potentially no fuel cost over its life (as for most RE) depends on the historical accident of whether global fuel prices were high or low in the previous year.

- **BPP does not reflect the nonfinancial benefits of RE generation.** For example, it takes no account of reductions in air pollution that result from substituting RE generation for coal-fired generation. As a result, the use of BPP as the benchmark for RE pricing undervalues the benefits of RE generation and results in levels of development below those that are socially optimal.

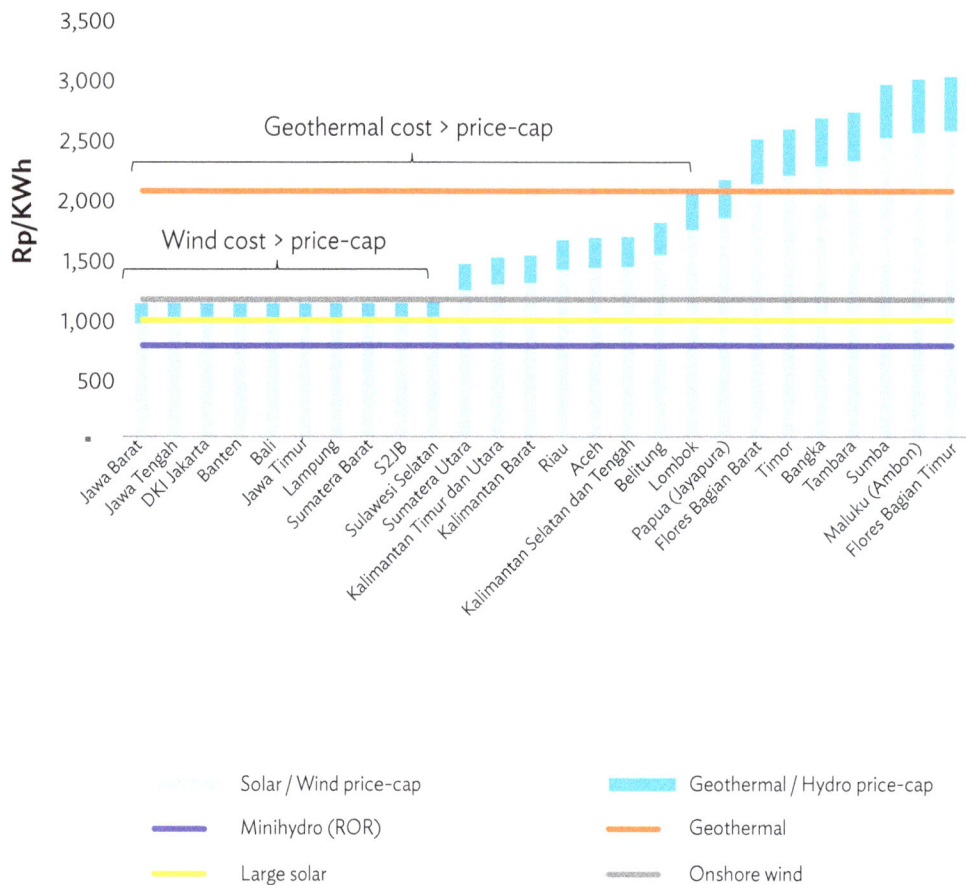

Figure 2:   Price–Cost Gap for Selected Renewable Energy Technologies

BPP = *biaya pokok produksi* (production cost), PT SMI = PT Sarana Multi Infrastruktur, ROR = Run of river, Rp = Indonesian rupiah.

Notes: Renewable energy technology costs are estimated from Ea Energy Analysis. 2017. *Technology Data for the Indonesian Power Sector*. National Energy Council, with the exception of geothermal where estimates are derived from GT Management. 2019. *Cost of Production from Geothermal Power Projects in Indonesia*. PT SMI. Funded by New Zealand Foreign Affairs and Trade Aid Programme. Geothermal/hydro price cap is higher of 100% of national average BPP or regional BPP. Solar/wind price cap is higher of 85% of national average BPP or regional BPP.

Source: ADB.

The estimates are generic in nature and do not fully reflect the impacts on RE project costs of some of the nonprice barriers previously identified, such as local content requirements, risks of delays, and restrictions on foreign investors. These are likely to push up costs above those shown in Figure 2 and, therefore, imply that solar and minihydro projects may require subsidies in some cases. The estimates also do not take account of the higher costs associated with projects in more remote locations, particularly in eastern Indonesia.

2.3   International Approaches to Delivering Subsidies

Internationally, most countries have chosen to pay a cost-recovering price for renewables purchases, implicitly compensating off-takers for any difference between the RE price and the cost of the alternative through a subsidy or increase in tariffs. The RE price may be set administratively, as a feed-in tariff (FIT), or through auction mechanisms designed to reveal the "true" cost of renewables. It may also be indirectly, through the imposition of a Renewable Portfolio Standard (RPS) on suppliers, requiring them to meet a given quota for RE purchases or pay a penalty—meaning suppliers are willing to pay an RE price up to the level of the penalty to comply with the RPS and avoid the penalty.

Within the Association of Southeast Asian Nations or ASEAN, approaches adopted include:

- FIT mechanisms in Thailand and Viet Nam, both of which attracted major waves of investment due to the FIT exceeding RE project costs. In Thailand, this led to the imposition of quotas followed by suspension of the FIT and, in Viet Nam, to downward revisions of the FIT. Viet Nam also applies an avoided cost tariff for minihydro, which again triggered large-scale investment and led to changes in the calculation methodology to reduce the avoided cost estimate, which was considered as being excessive in retrospect.

- Competitive auctions for large-scale solar projects in Malaysia. These have also proven very successful in attracting investors, much more so than the use of a FIT mechanism for biomass, where prices were seen as too low relative to project costs.

- A FIT mechanism in the Philippines, which has now been replaced by the adoption of an RPS with accompanying trading of obligations between suppliers.

Indonesia has previously applied both a FIT based on estimated RE technology costs and an auction mechanism before settling on the current BPP-linked price mechanism.

It is worth noting that in many countries the costs of RE technologies are now coming close to or even below the costs of fossil fuel generation. This has led to support mechanisms increasingly being related to de-risking, by providing guaranteed prices and/or other support such as land under long-term contracts with credit-worthy off-takers.[4] For Indonesia, an increased emphasis on de-risking may help deliver similar results where rapid deployments of many RE technologies is possible with little or no explicit subsidy if it becomes easier for developers to obtain long-term PPAs with guaranteed revenues.

Whatever support mechanism is adopted, there remains a need to recover any difference between the RE price and the cost of the alternative supply source. The two main mechanisms for doing so internationally are:

- **Include the higher RE purchase costs directly into customer tariffs.** Where tariffs are regulated, this can be done directly. Where they are unregulated, in a competitive market, then suppliers will decide by how much to mark up their tariffs to recover the additional costs.

- **In competitive markets, recover the difference in costs through a levy on electricity sales**, usually on network charges which cannot be bypassed and so must be paid by all suppliers and their customers, and then redistribute to RE buyers. For regulated monopoly markets, such as in Indonesia, this mechanism

---

[4]   For example, the third round of offshore wind auctions in Great Britain, held in September 2019, attracted winning bids of 5.0 US cents per kilowatt-hour (USC/kWh)–5.5 USC/kWh compared to a forecast 'reference' price of 6.5 USC/kWh. The implication is that RE project developers are winning to pay a premium (a 'negative subsidy') to obtain long-term contracts.

is unnecessary—it leads to the same outcome as directly including the higher costs in regulated tariffs while adding complexity and cost to the process.

Both these mechanisms ultimately end up passing the higher costs of RE purchases into customer tariffs. However, this has been ruled out in Indonesia due to the sensitivities around tariff increases and the impacts on affordability.

Given this scenario, the **most feasible option in Indonesia for closing the gap between RE costs and PLN's purchase price is some form of subsidy funded from the state budget ("budget subsidy")**—whether provided directly to RE project owners to reduce the price they need to charge to PLN, or to PLN to compensate for the higher cost of purchasing from renewables.

2.4   Options for Delivering Budget Subsidies in Indonesia

There are various means for delivering budget subsidies to RE projects in Indonesia:

- Subsidized loans to bring down the cost of financing for RE projects.

- Fiscal incentives, such as tax breaks, to bring down the cost of RE projects.

- Indirect subsidies, such as provision of free land and/or supporting facilities and infrastructure, to also bring down the cost of RE projects.

- Direct cash subsidies to RE project owners, allowing them to charge lower prices to PLN and still remain viable.

- Direct cash subsidies to PLN to close the gap between its avoided costs (cost savings) and RE purchase prices, thereby enabling it to pay higher prices while not increasing retail tariffs.

An assessment of these options is provided as follows:

- **Subsidized loans** could help reduce costs but are unlikely to bring this down to levels consistent with PLN's willingness-to-pay. As an illustration, the premium of A-rated corporate 10-year bonds (used as a proxy for the costs of financing a RE project selling to PLN) over government bonds is currently around 360 basis points. Removing this premium through an interest rate subsidy would reduce the estimated cost of a geothermal project by 16%, from Rp2,068/kilowatt-hour (kWh) to Rp1,733/kWh. This remains above PLN's average generation BPP and so does not resolve the cost–price difference. An assessment by *Badan Kebijakan Fiskal* (Fiscal Policy Agency) came to similar conclusions that a reduction in interest rates of 3% leads to payback periods falling only slightly from 10.1 years to 9.1 years and prices falling by only 13%.[5] Moreover, since subsidized loans are provided prior to plant operation, there is a moral hazard risk that loans are made to developers that are unable to complete their projects.

- **Fiscal incentives,** such as tax allowances and tax holidays. These are already in place in Indonesia. However, these incentives have proven insufficient to trigger the scale of investment that is required. Since 2016, tax allowances have been granted to seven projects and tax holidays to nine projects, with a

---

[5]   Badan Kebijakan Fiskal. 2018. *Kajian Analisis Dampak Insentif Fiskal terhadap Investasi dan Harga Jual Listrik dari Energi Terbarukan.*

combined investment of Rp63.2 trillion ($4.5 billion) or an annual average of Rp16 trillion ($1.1 billion).[6] While impressive, this investment level falls noticeably short of the $4 billion required annually over the next 10 years under current targets.[7]

- **Indirect subsidies** face a number of challenges. There is the question of whether these will be sufficient to close the cost–price gap. For a geothermal project, for example, the costs of land acquisition, roads, and other infrastructure, items which might be funded by government, make up only around 3% of total capital expenditures and so the contribution would be relatively small. The government's recent geothermal exploration risk mitigation facilities can help reduce exploration risks and costs for geothermal developers. While this will result in lower costs for any given project, the oversight and administration required for such programs make them difficult to scale up in a timely manner.

- **Direct subsidies to individual RE projects** from the government budget would raise concerns over the potential misuse of funds. The members of the House of Representatives (DPR) have previously ruled out such mechanisms due to the difficulties of reconciling oversight of such subsidies with the need to commit to multiyear payments to RE projects. Introducing these mechanisms may require new primary legislation to allow for such commitments.

- **Direct subsidies to PLN** have a number of key advantages. They can be delivered at the scale needed to deliver RE generation expansion at the levels needed to meet Indonesia's targets. They would be delivered through the annual budgeting process, enabling ESDM, MOF, and DPR to maintain oversight over their levels and use, and they can be introduced under existing primary legislation.

A summary comparison is shown in Table 2. Summary Assessment of Options for Subsidies. Based on this, **we recommend that any subsidy should be delivered in the form of direct budget payments to PLN** to cover the gap between its cost savings from RE power purchases and the costs of these purchases.

---

[6] Ministry of Finance data.
[7] Calculated assuming an annual average addition of: Geothermal – 460 MW, Large hydro – 800 MW, Minihydro – 155 MW, Solar – 90 MW, Wind – 85 MW, and Biomass + Waste – 80 MW [RUPTL 2019–2028]. Investment costs are estimated at: Geothermal - $3,500/kW; Large hydro - $2,000/kW; Minihydro - $2,600/kW; Solar (on-grid) - $830/kW; Wind (onshore) - $1,500/kW; and Biomass - $1,700/kW [Ea Analysis (2017)].

Table 2:   Summary Assessment of Options for Subsidies

Subsidy Option	Legal and Institutional Feasibility	Effectiveness in Closing Price–Cost Gap[a]	Potential for Scaling-Up
Subsidized loans	HIGH. Already implemented (through PT SMI)	LOW. Unlikely to reduce costs sufficiently to bring these within price caps	MEDIUM. Limited capacity to manage much larger loan portfolio
Fiscal incentives	HIGH. Already implemented	LOW. Current fiscal incentives appear inadequate to deliver significant RE investments.	LOW. Difficult to increase current incentives
Indirect subsidies	VARIES. Depending on the particular mechanism	VARIES. Costs covered by indirect subsidies may be only a small part of the total project cost depending on the particular mechanism	MEDIUM. Development, administration, and oversight of such programs by government take time and resources, adding another step to the development process
Direct subsidies to individual RE projects	LOW. Legally uncertain given concerns over loss of oversight	HIGH. Subsidy can be set to close entirety of gap	HIGH. Subsidy payments are only limited by fiscal concerns
Direct subsidies to PLN	MEDIUM. Allowed under existing primary legislation for subsidies to state-owned enterprises but regulatory challenges exist	HIGH. Subsidy can be set to close entirety of gap	HIGH. Subsidy payments are only limited by fiscal concerns

PLN = *Perusahaan Listrik Negara* (State Electricity Company), PT SMI = PT Sarana Multi Infrastruktur, RE = renewable energy.

[a] For geothermal projects.

Source: ADB.

3. Budget Subsidy Mechanism

3.1 Legal Basis

The proposed budget subsidy to PLN is grounded in Law 19/2003 on State-Owned Enterprises and Law 30/2007 on Energy:

- Law 19/2003 stipulates, in Article 66, that the Government must compensate a state-owned enterprise (SOE) for its costs plus margin whenever Government imposes a public service obligation (PSO) on the SOE that is financially infeasible (i.e., the cost of delivering the PSO is greater than the revenue received).

- Law 30/2007 stipulates, in Article 21, that the Government may offer facilities and/or capital, tax, or fiscal incentives for renewable energy development until renewables become economically competitive.

The legal basis for the subsidy, therefore, would derive from the government under Law 30/2007 imposing a PSO on PLN for the procurement of renewable energy. The incremental costs of this PSO would then be recovered under the compensation mechanisms under Law 19/2003.

3.2 Governance

Governance of the subsidy will follow that for the existing electricity tariff subsidy paid to PLN to cover the gap between its costs and revenues from subsidized customers. In brief, this would involve the following steps:

- PLN calculates the annual estimated subsidy requirement, following the methodology described.

- This estimate is verified by ESDM and MOF, and included in the draft State Budget (APBN).

- The subsidy estimate is reviewed by DPR and the final value included in the approved APBN.

- Payments are made in arrears on a monthly basis, for 95% of the total estimated subsidy for the year.

- Following the end of the fiscal year, an audit is conducted to determine the final subsidies payable relative to estimated levels, given actual levels of renewable energy production. Where the actual value is below the estimate included in the APBN, the difference is recovered from the remaining unpaid 5% of the subsidy. Where the actual value is higher, the difference is carried forward to the next year's budget.

3.3   Budgeting for Subsidies

The annual subsidy estimate represents the difference between the projected costs of PLN purchases or own generation from eligible RE projects in each year and PLN's resulting cost savings from the avoidance of fossil fuel generation costs. This is calculated according to the following formula:

$$BDGT_y = \left(1 + m_y\right) \times \sum_{i=1} \left(SUBS_{i,y} \times SALE_{i,y}\right)$$

$$SUBS_{i,y} = max\left[0, \left(REPC_{i,y} - COST_{i,0}\right)\right]$$

Where:

$BDGT_y$ Annual budget subsidy estimate in current year y (Rp)

m_y Allowed PLN margin in current year y (%)

$SUBS_{i,y}$ Annual subsidy required for RE project i in current year y (Rp/kWh)

$SALE_{i,y}$ Forecast sales to PLN for RE project i in current year y (kWh)

$REPC_{i,y}$ RE project cost for project i in current year y (Rp/kWh)

$COST_{i,0}$ Financial cost saving (avoided cost) for PLN from RE project i in year of PPA signature 0 (Rp/kWh)

As the purchase price and cost savings may differ by RE project, the calculation must be performed separately for each project and then summed to obtain total estimated subsidies for the year. RE projects for which PPAs have been signed but which are not in operation at the start of each year will be included in the calculation starting from the expected commercial operations date for the remainder of the year.

As with the tariff subsidy, this calculation includes provision for a margin to PLN for providing the PSO (in this case, the expansion of RE generation) following the requirements of Law 19/2003. This margin can be interpreted as a payment to PLN to compensate for additional administrative costs resulting from the PSO, as the renewable subsidy calculation only captures the difference between the RE project's own costs and PLN's avoided generation costs. This is expected to be a relatively small amount on the order of 0.5%, intended to cover PLN's time value of money for the period between when it pays for RE power and when it receives the subsidy payment from government, plus the cost of administration.

This would be noticeably lower than the margin applied under the tariff subsidy, which is currently set at 7%. The margin applied to the tariff subsidy is the result of negotiations with DPR to ensure PLN is compensated for the difference between its revenue from and cost of service to subsidized tariff classes. On the other hand, the RE subsidy is only intended to compensate PLN for the direct cost difference between two different sources of supply.

It will be important to ensure that MOF does not pay excessive subsidies for renewables projects relative to both the:

- costs of the RE project, so as to avoid subsidies that exceed those needed to mobilize new investments, and

- the socioeconomic benefit of the project to Indonesia, so as to avoid projects whose total costs exceed their social benefits.

RE project costs should be determined by auction, but auctions are not suitable for all RE technologies. In cases where auctions are not used, the price should be set by government. The following sections discuss how each element of the proposed framework is to be determined:

- the RE project price when auctions are used,

- the RE project price when auctions are not appropriate,

- the economic value cap, and

- PLN's financial avoided cost.

3.4   Auction-Based Renewable Energy Project Prices

$$REPC_{i,y} = min\left(PPAS_{i,y}, ECON_{t,g,0}\right)$$

Where:

$PPAS_{i,y}$ Signed PPA average price for RE project i in current year y (Rp/kWh)

$ECON_{t,g,}$ Economic benefits for RE technology t on PLN grid g in year of PPA signature 0 (Rp/kWh)

The price for an RE project should be established through a competitive tendering process whenever possible. In most cases, this would be expected to be a reverse auction, under which developers are awarded projects based on the lowest bid price. Projects then are ranked from lowest to highest price and selected in order until the target procurement quantity is reached.

The benefits of relying on auctions rather than production cost estimates can be large.[8] Reverse auctions are a well-established procurement process for renewables, particularly for technologies like solar photovoltaic (PV) which are more generic in nature, less site-dependent, and the resource is relatively well understood. The lowest solar PV project prices globally have been established through reverse auctions, often in conjunction with indirect subsidies such as the provision of land.

Reverse auctions are less suitable for projects where the selection of sites is very limited, such as some hydropower tenders, and for technologies where resources are uncertain and need to be proven in advance, such as geothermal projects, where licence awards are based on exploration offers which may or may not lead to an actual project.

Reverse auctions can be conducted as a two-step process: first, prequalification based on fulfilment of financial and administrative criteria together with a satisfactory track record of past projects. All project documents such as the PPA should be provided at this stage, and the criteria should be clear from the outset. Any firm should be allowed to submit an expression of interest. In effect, the off-taker (PLN) should be prepared to contract with any prequalified bidder. The prequalified firms then compete purely on price.

---

[8]   For example, it is estimated that relying on generic production cost estimates to calculate subsidies for geothermal projects in Java-Bali, rather than using an auction process which reveals the 'true' costs of each individual project, could increase subsidies by as much as 170%, due to subsidies being higher than those needed to incentivize investment in lower-cost projects. Illustrative calculations are set out in the appendixes.

Prequalified firms are then invited to submit price proposals, and the lowest price offered by a technically, administratively, and financially compliant bid should be accepted. The winning firm should not be subsequently subjected to negotiations with a view toward further reducing the price, since bidders will take into account in pricing the uncertainty and associated risks arising from imposition of post-tender price negotiations.

The direct selection method currently applied by PLN for most technologies as stipulated by PERMEN ESDM 50/2017 does not facilitate sufficient competition to be used to establish prices. It is closed to firms that have not registered in advance for the generic *Daftar Penyedia Terseleksi* (List of Selected Providers). The list is opened only once every 3 years on a general basis, but developers are typically interested to go through prequalification if they know what specific project is being offered. Moreover, under prevailing regulation, direct selection is applied for hydro, biomass, and biogas projects, which are typically tied to a particular resource or feedstock; this makes a truly open competition difficult if not impossible.

3.5 Renewable Energy Project Prices Using Cost Estimation

$$PPAP_{i,y} = min\left(PPAS_{i,y}, COST_{t,g,0}, ECON_{t,g,0}\right)$$

Where:

$COST_{t,g,0}$ Cost-cap for RE technology t on PLN grid g in year of PPA signature 0 (Rp/kWh)

Reverse auctions are not suitable for all projects or technologies. Consider for example geothermal, for which working areas are awarded on the basis of the best exploration offer. Geothermal developers formulate their exploration offers on the basis that if successful, they will be able to develop the power plant. Subjecting geothermal prospects that have been proven to another round of tendering for the right to build a power plant will adversely affect willingness of developers to take on exploration risk, and further delay development.

One option in these cases is for the government to establish a production cost model that determines a tariff based on factors outside of the control of the developer. When a developer comes forward with a project, the price at which project output will be purchased is set by the model based on the specific characteristics of the project.

To develop such a model, the government must compile information about input costs as a function of parameters outside of the control of the bidder, such as project size and resource quality. The government must also determine reasonable returns and project financial structure. Every developer has its return requirement, and every project is structured financially in its own way. The government should not change assumptions for each project or otherwise try to guess the particular return requirement and financial structure for each developer and project, but rather should select returns and financial structures that are reasonable, consistent with typical good practice in the industry, and can be justified for entire classes of projects.

The length of time between the signing of a PPA and commissioning should also be taken into account. For some types of technologies, such as grid-connected PV, this will typically be 2 years or less. But for other technologies such as geothermal, commissioning can take place several years after a PPA is signed; the model should take into account inflation, time value of money, and interest during construction typically associated with projects of that size and nature, when calculating the price the sponsor will receive.

Each RE technology not procured through reverse auctions will need a different production cost model. Ideally, the model will use recent information on the costs of similar projects located in Indonesia (with international costs being substituted where Indonesia-specific data is not available). As part of its 2018 report on renewable financial incentives, the *Badan Kebijakan Fiskal* (Fiscal Policy Agency) developed such models for small hydro and solar PV projects. A separate production cost model has also been developed for geothermal, and is applied in this report to estimate future subsidy payments. This model was prepared by Jim Randle and Jim Lawless for the Directorate General of New and Renewable Energy and Energy Conservations under funding from the New Zealand Ministry of Foreign Affairs and Trade, and subsequently updated by the authors for use by PT Sarana Multi Infrastruktur and others.

These production cost models should be owned and maintained by MOF and prepared in consultation with ESDM to avoid potential conflict of interest such that PLN would have to inflate production cost estimates so as to earn higher subsidies. Nonconfidential inputs and assumptions used should be published on the MOF website for transparency and provide confidence to potential investors. The underlying data should be reviewed at least on an annual basis and updated when additional information has become available. It is imperative that the government establish only one set of models for this purpose to provide clarity and certainty to all stakeholders.

Setting the price in this manner would relieve PLN and the developer the burden of negotiations over PPA prices. The production cost estimate would become the de facto PPA price, as PLN would receive a subsidy up to this amount and so should be willing to enter into a PPA with a price equal to this estimate. This could greatly accelerate contracting and, provided that the model is made publicly available, developers would have certainty regarding how the price of their project output will be set, reducing their risks and easing financing of investments.

While it is possible to use the production cost model to set a ceiling for PLN negotiations with the project sponsor, this would be counterproductive as it would delay implementation of projects for years and would force developers to seek higher returns to compensate for the risk of negotiations with PLN. While it might be possible for PLN to squeeze out some savings through negotiations, the purpose of this framework is to accelerate renewable uptake. Moreover, the government prepares the production cost model and is therefore assured that the resulting prices are reasonable and based on industry good practice. Finally, the recommended framework should also be applied to RE projects that PLN develops on its own. If the framework were used to establish a ceiling price for negotiations rather than the actual PPA price, who would negotiate with PLN for the RE projects it develops on its own?

While this approach has similarities to feed-in tariffs (FITs), it is distinguished by its much greater level of granularity in the cost estimates. Generally, FITs can only be differentiated by a small number of parameters, such as size and location. A production cost model which is separately applied by technology and PLN system reduces the risk that the cost estimate and the actual costs of a specific project are very different and, thereby, reduces (but does not eliminate entirely) the likelihood that a developer earns windfall profits, i.e. that a particularly efficient or capable developer may be able to earn more than the amount it would have been willing to accept to develop the project. On the other hand, the developer continues to bear the risks that it is best placed to manage, and in some cases due to bad luck or poor management, may earn returns lower than those assumed in the production cost model.

FITs may be appropriate for technologies like wind that are not amenable to reverse auctions and not subject to the extent of resource risks that characterizes geothermal. Given the many resource parameters that are outside the control of geothermal developers control and the high sensitivity of a project's financial viability to these parameters, a production cost model is the only way to adequately capture these factors. For example, the previously cited model developed by Randle and Lawless considers ten parameters to characterize each project. FITs on the other hand are suitable when there are only two or three parameters outside the developers control, e.g., location (to capture system avoided cost) and size (to capture economies of scale).

Figure 3 shows how a production cost model can maximize development while minimizing payments to generators when compared to a FIT. In this example, the power purchase price determined by the FIT or the production cost model is a function of only one parameter, installed capacity. In practice, it is impractical to define a FIT in terms of more than three parameters, whereas a production cost model can incorporate as many parameters as the government deems appropriate. The "minimal acceptable price contour" represents the minimum price a developer would accept for any particular level of installed capacity. The production cost model is designed to approximate this curve. In this example, acceptable price declines with installed capacity, indicating that the technology is characterized by economies of scale.

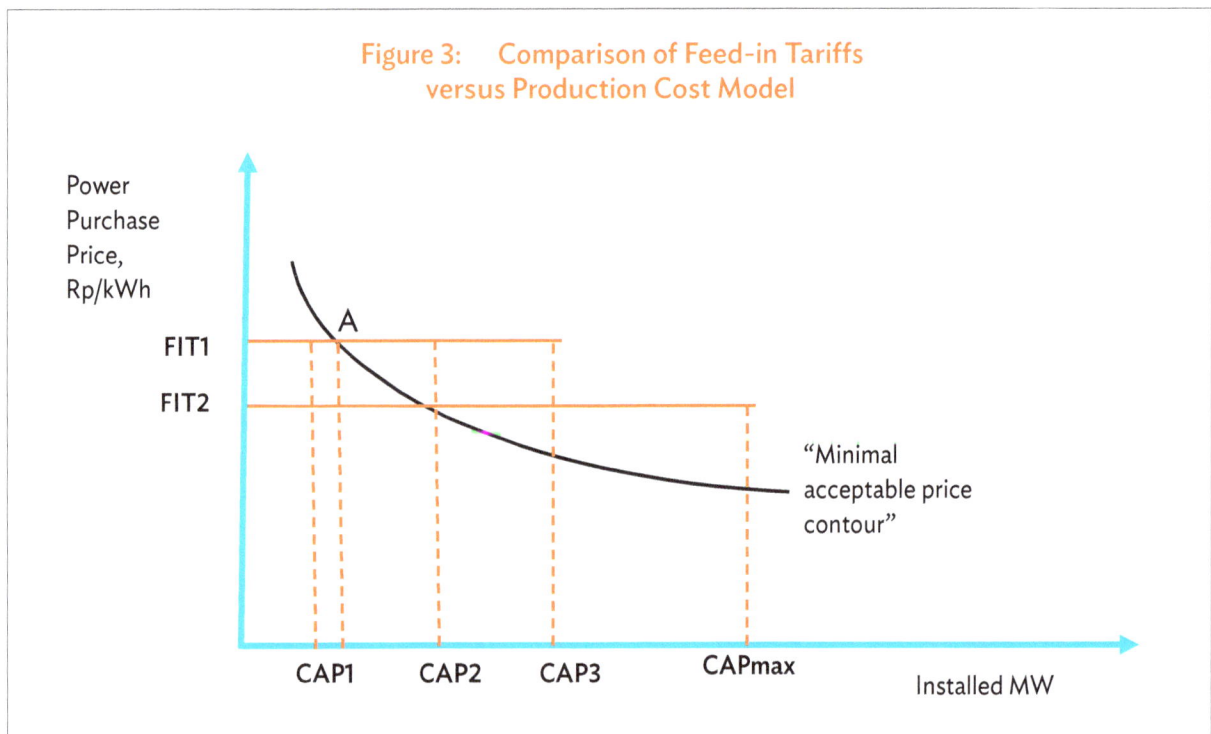

Figure 3:   Comparison of Feed-in Tariffs versus Production Cost Model

FIT = feed-in tariff, MW = megawatt, Rp/kWh = rupiah per kilowatt-hour.

Source: ADB.

In this simplified example, only two levels of feed-in tariff are considered: FIT1 and FIT2, and the cut-off for the power purchase price to shift from FIT1 to FIT2 depends on the installed capacity of the project.

- If the cut-off between FIT1 and FIT2 were set at CAP1, then only plants CAP2 or larger would be built, and plants larger than CAP2 would be overpaid relative to the minimum payment they would be willing to accept.

- If the cut-off were set at CAP2, then only plants larger than A would be built. Plants between A and CAP2, and larger than CAP2 would be overpaid. Only plants of size CAP2 would receive the minimal acceptable price.

- If the cut-off were set at CAP3, then only plants larger than A would be built, and all plants would be overpaid.

In reality, each developer has its own minimal acceptable price contour. In establishing its production cost model, the government does not intend to define a pricing regime acceptable to all potential developers or tailored to each individual developer. Rather, it only aims to estimate a single minimal acceptable price contour that is reasonable, justifiable, and consistent with industry good practice. This can still result in overpayment or underpayment for any particular developer or project, but it greatly minimizes this risk relative to a FIT regime.

3.6   Economic Value Ceiling

Regardless of how the RE project's cost is determined, the subsidy should be capped by the economic value of the project. This ensures that the government does not subsidize projects in excess of their value to the country.

The economic benefits of RE projects are defined as the sum of PLN's economic avoided costs and externalities, including reduced pollution, reduced carbon emissions, and contributions to local economic development. The avoided costs are estimated using the proxy generator method, wherein the most likely alternative generator is identified and its costs estimated. Fuel costs for the proxy generator are based on expected or forecasted international market prices to reflect their economic value, not the price that PLN would actually pay for the fuel.

The proxy generator will differ from system to system, between RE technologies depending on their operating mode and with the supply–demand balance (which determines whether or not a new generator would need to be constructed to meet demand growth). It will, therefore, be necessary to calculate a separate economic benefit for each RE technology covered by the subsidy mechanism and for each PLN grid.

A listing of economic benefits and a sample calculation is contained in the appendices. Figure 4 provides a summary of this calculation for a sample geothermal project located in Java–Bali and for which the alternative generator is assumed to be a new coal-fired power plant. As can be seen, these economic benefits are estimated to exceed the costs of power from a generic geothermal plant.

Figure 4: Economic Benefits Example (Java–Bali Geothermal Replacing Coal)

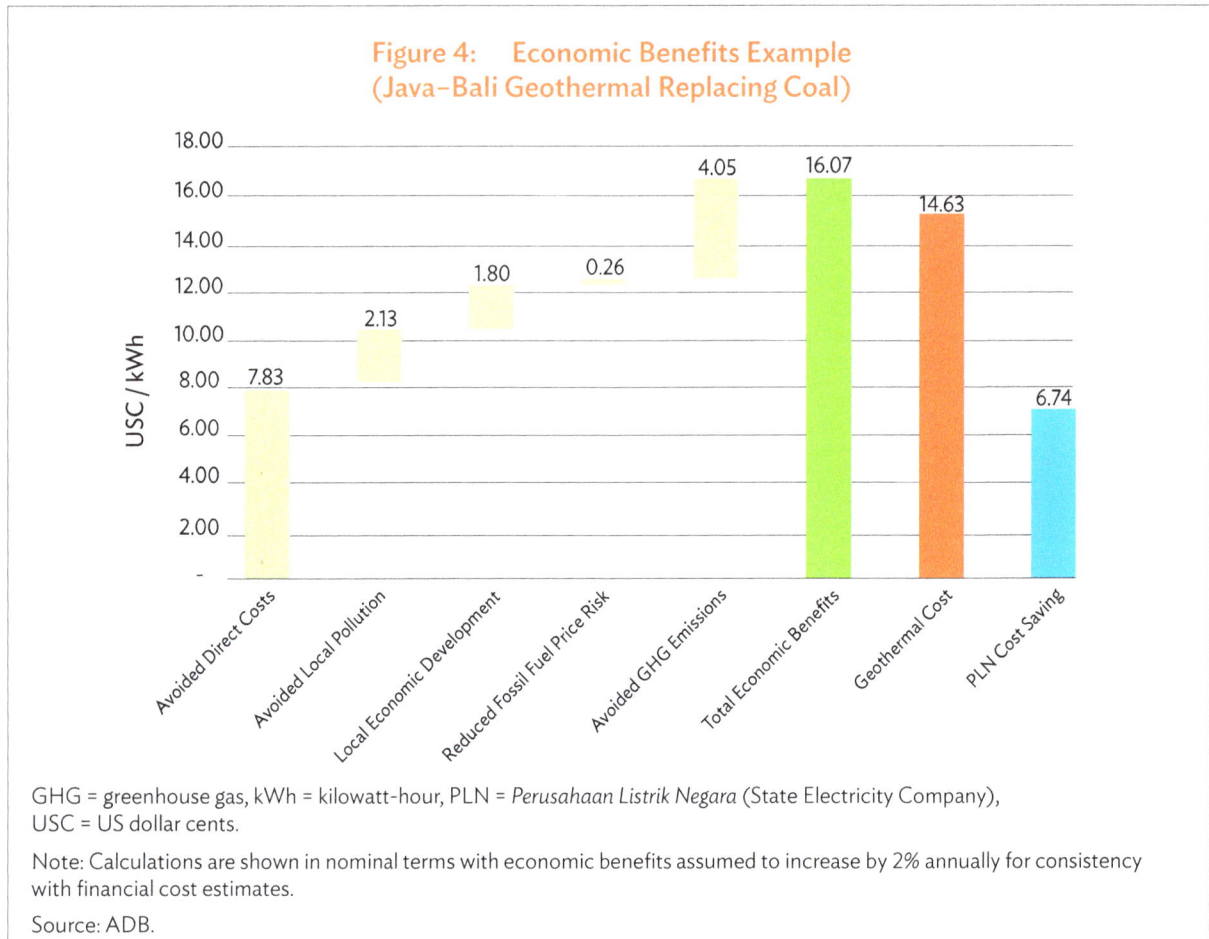

GHG = greenhouse gas, kWh = kilowatt-hour, PLN = *Perusahaan Listrik Negara* (State Electricity Company), USC = US dollar cents.

Note: Calculations are shown in nominal terms with economic benefits assumed to increase by 2% annually for consistency with financial cost estimates.

Source: ADB.

PLN, as the best-informed entity with regards to its own costs and investment plans by grid, will propose the economic benefits calculation. This should use the same assumptions on the economic costs and the technological characteristics of alternative technologies as are applied in the preparation of the Electricity Supply Business Plan (RUPTL). Assumptions and inputs not used in the RUPTL such as the social discount rate; conversions from financial to economic costs, if required; and the economic costs of pollution will be specified by MOF, and will be updated as required. Conflicts of interest are less of a concern in this instance than with the estimation of production costs, as the estimated economic benefits form a cap on PPA prices used for subsidy calculations but do not otherwise determine the subsidy received.

All assumptions should be made publicly available on the PLN website along with the model used for the calculation and accompanying explanatory notes. This is essential for RE project developers, MOF, and DPR to have confidence in the resulting estimates used in the calculation of subsidy requirements. Many of the valuations used are not directly observable and depend on assumptions as to, for example, the level of exposure to health impacts of pollution from a generic new coal plant and the value of the resulting loss of life. These values inherently require significant judgement.[9] Publishing them enables stakeholders to understand the basis for calculations and, if necessary, challenge these and propose alternatives which they consider better reflect

---

[9] An example is provided by the estimation of avoided local pollution externalities. In the example provided here, these are valued at 2.13 USC/kWh, derived from recent IMF estimates. By contrast, the earlier World Bank/Asian Development Bank (ADB) report on "Unlocking Indonesia's Geothermal Potential" valued these impacts as near-zero on the assumptions that air pollution impacts are very limited with modern equipment and that the values of their impacts when spread across the wider population are negligible.

the magnitude and values of externalities. The calculation of economic benefits will be updated annually to reflect the changing supply–demand balance on each grid and changes in the costs of alternative supply sources, notably fuel prices.

The model, input data, and assumptions used will be reviewed and approved by both MOF and ESDM. Review by MOF is essential given the use of the model for subsidy calculations. Review by ESDM is required to assure that data and assumptions are consistent with those used for power sector planning purposes, and that these are reasonable. The model will be approved when first prepared, and again whenever any significant updates or amendments are made. Input data and assumptions will be approved annually. Minimum requirements for the model's outputs and the review and approvals process will be stipulated in the implementing regulations issued by MOF.

3.7   Financial Avoided Cost

PLN's financial cost savings, or financial avoided cost, will be calculated using the same proxy generator method for calculating economic benefits, but applying financial costs rather than economic costs. The key differences between these concepts, as explained in the appendices, lie in the assumed discount rate or cost of financing applied and in the assumed fuel prices faced. This can result in a financial cost saving above or below the economic avoided costs, as shown in Figure 4.

While the proxy generator is an imperfect estimate of PLN's financial cost savings, it has the advantages of being relatively simple to define and calculate while still being a better measure than the current use of BPP.[10] It also simplifies the calculation process by being aligned with the calculation of avoided costs under the economic benefits calculation. A separate financial cost savings calculation will be needed for each RE technology and each PLN grid, as the proxy generator will differ in each case.

As with the calculation of economic benefits, the calculation of financial cost savings will be proposed by PLN, using the same input and assumptions as in the RUPTL as far as possible, and using a model, input data, and assumptions reviewed and approved by both MOF and ESDM. The assumptions and model will be published on PLN and government websites. The cost savings estimates will be updated annually.

The PLN financial avoided cost used to calculate subsidies for individual RE projects will be fixed at the time of PPA signature. The use of this "fixed" cost saving rather than a "floating" cost saving updated each year reflects PLN's risk exposure. For example, assume an increase in coal prices equivalent to $0.01/kWh relative to the forecast for calculating the financial avoided cost. From PLN's perspective, the following applies:

- The cost of coal-fired generation has risen by $0.01/kWh. In principle, this is offset by an increase in retail tariffs under the tariff indexation mechanism (if the current suspension is lifted) so that PLN's net revenues are unchanged.

---

[10]   There are various methods available to determine a utility's avoided cost of generation. The most sophisticated is the "differential revenue requirements" methodology, which entails execution of a capacity expansion planning model with and without a particular project to determine avoided costs. However, neither PLN nor the government currently has the capability to conduct this sort of analysis, so the simpler and more intuitive proxy unit method is proposed.

- The costs saved by using existing RE generation rather than coal-fired generation are unchanged, given that the output of these generators cannot be adjusted to replace coal-fired generation.[11] Consequently, the required subsidy is also unchanged.

- The costs saved from using new RE generation have risen and, therefore, future subsidy needs associated with new RE projects are reduced accordingly.

3.8   Auditing Subsidies

Under current arrangements, the tariff subsidy paid to PLN is finalized annually based on an audit of

- PLN's historical accounting cost of supply at different voltage levels over the prior year as represented by BPP,

- the number of units actually sold to subsidized tariff classes, and

- the interim tariff subsidy payments received by PLN over the past year.

Similarly, the proposed renewable subsidy paid to PLN will be finalized based on an audit of the actual number of units each subsidized RE project sold to PLN and the interim renewable subsidy payments received by PLN over the past year (this would include correction for differences between estimates of prices for new RE projects expected to commission during the year, and the prices that are ultimately adopted).

However, one key difference between the existing tariff subsidy and the proposed RE subsidy is that because BPP is an actual historical cost, it can also be audited.[12] In contrast, PLN's financial avoided cost and any estimated RE production cost used to determine RE subsidies are not actual historical costs, and hence cannot be audited in the traditional sense. Over time, actual costs are likely to deviate from these forecasts and estimates, e.g., the actual trajectory of future fuel costs is likely to differ from the fuel cost forecasts used to estimate PLN's financial avoided costs.

For the proposed RE subsidy framework to function, the government must be willing and able to define and implement a subsidy scheme based on forecasts and estimates fixed at the time of PPA signature. Auditing by the *Badan Pemeriksa Keuangan* (State Audit Agency) will follow whatever framework is adopted by the government. The role of any auditor, including the State Audit Agency, is not to set the standards of auditability for any project, but rather to audit whether projects have been developed in accordance with the laws and regulations put in place by the government.

Transparency and public participation will help ensure the integrity and reasonableness of the PLN avoided cost and RE production cost calculations. The following measures are proposed to achieve this:

- PLN will propose avoided costs annually consistent with the most recent approved RUPTL, and these avoided costs will be reviewed and approved by ESDM and MOF.

---

[11]   This assumes that no change in RE generation levels is possible, given that they are already expected to run at maximum possible output given their zero short-run costs.

[12]   It is understood that one of the reasons BPP was used as the basis for establishing tariffs payable to RE projects initially under PERMEN ESDM 12/2017 and subsequently under PERMEN ESDM 50/2017 is that as a historical cost that can be audited, it is unambiguous and difficult to manipulate.

- MOF will prepare and maintain the RE production cost models used to set or cap payments to RE projects in the absence of competitive tendering.

- All assumptions and calculations will be available to the public on websites and other media to enhance transparency.

- The public should be given a reasonable period to review and comment on the proposed assumptions and models before they are approved. Inputs received from the public should be available for public inspection; in making final approvals, ESDM and MOF should respond to these comments.

3.9   Regulatory Requirements

3.9.1   Implementing Regulations

The budget subsidy mechanism will be implemented through a regulation issued by MOF. This will specify the process for estimating, approving, paying, and auditing subsidies, including the principles for the calculation of economic benefits and PLN's financial cost savings and the process for reviewing and approving these calculations. The actual calculations will be undertaken by PLN itself, given its access to the necessary information and to ensure consistency of assumptions. The regulation will also specify how MOF will validate the proposed RE supply price for use in the subsidy calculation, but will not, itself, stipulate how PPA prices are set as this is falls under ESDM's authority.

3.9.2   Amendments to ESDM Regulations

As described earlier, existing ESDM regulations prevent PLN from paying RE projects a price in excess of the BPP-linked cap. This raises a challenge for the implementation of the proposed subsidy mechanism in that, even if PLN receives a subsidy for the difference between the RE project's costs and PLN's own cost savings, it is unable to increase the purchase price accordingly.

Addressing this requires one of two regulatory changes:
- Ideally, **issuing a new regulation to replace the existing RE pricing regulations** (PERMEN ESDM 50/2017 and 53/2018). The new regulation should include the following provisions:
  - Retaining a price cap linked to PLN's avoided costs.
  - Allowing for the price cap to be relaxed where a subsidy is provided for the difference between the PPA price and avoided costs.
  - Replacing the use of BPP in setting avoiding costs with a financial cost savings calculation, aligned with the applied by MOF for the subsidy calculation.
  - Providing for the use of competitive tenders (specifically, reverse auctions) to procure RE projects in place of the direct negotiation model.

- Alternatively, **if existing RE pricing regulations are retained, a new MOF regulation will need to be issued covering subsidy payments from PLN to RE project owners**. In this case, PLN needs to be allowed to enter into separate but accompanying contracts alongside the PPA itself. The contracts will pay an additional fee to the RE project for "environmental services." This additional fee will be equal to the budget subsidy allocated for that project, adjusted for any differences between the value of PLN's financial cost savings calculated and the BPP value used in establishing the PPA price.[13]

- The first option is clearly preferable. The second option is more complex and less transparent. It creates conflicting and overlapping regulations with, for example, both MOF and ESDM issuing regulations requiring the estimation of PLN's avoided costs, but with these being calculated on a different basis under each regulation. There may also be potential legal uncertainties if PLN is being authorized to enter into contracts with RE project developers by an MOF regulation, even if such contracts are not paying for generated energy and provision of capacity but for other services.

It is understood that a new presidential regulation that will establish a new renewable energy pricing and development regime is currently being prepared. The recommendations of this report may be considered for the drafting of this new regulation.

3.10 Geothermal Subsidy Estimates

Two scenarios were prepared to estimate future subsidy payments to new geothermal projects under the recommended scheme. The first scenario utilized the geothermal capacity additions planned in PLN's RUPTL 2019–2028. Given that over the past 15 years PLN has never achieved its geothermal capacity expansion plan (Figure 5), a second scenario was prepared that excluded all geothermal capacity additions in the RUPTL 2019-2028 for which prospects had not been specifically identified. In addition, the capacity for named prospects that have not yet been proven was adjusted to the estimates presented in an earlier study of geothermal potential in Indonesia.[14] Both scenarios considered only capacity additions for Java and Sumatera, as these systems represent 83% of all geothermal capacity additions planned in the RUPTL 2019–2028, and are the systems where geothermal is most likely to require subsidies.

For both scenarios, a detailed geothermal production cost model was applied to estimate the financial cost of supply from each project.[15] PLN's financial avoided costs were estimated based on plants identified in the RUPTL 2019–2028, and economic cost ceilings established based on analysis from other sources.

The RUPTL scenario results in an annual geothermal subsidy payment of Rp17.9 trillion for 3,786 MW of new capacity in 2028. In contrast, by 2028, the adjusted scenario results in an annual subsidy payment of Rp6.3 trillion for 1,546 MW of new capacity. These calculations are presented in Appendix 2.

These subsidy estimates may strike some observers as high, especially given estimates of Indonesia's geothermal potential. However, these estimates, such as the 11,999 MW of resources and 17,546 MW of reserves reported in the *Rencana Umum Energi Nasional* (RUEN, National Energy Plan) covering the period 2015 to 2050,[16] are highly speculative and have not been assessed in terms of commercial or economic viability. These speculative estimates of Indonesia's geothermal potential have consequently resulted in overly optimistic expectations

---

[13] This adjustment is necessary to ensure that PLN is not disadvantaged. If BPP is higher, for example, than the financial cost saving then the subsidy calculated using BPP would be less than the value of the savings from the RE project to PLN.

[14] Castlerock Consulting. 2010. *Phase 1 Report: Review and Analysis of Prevailing Geothermal Policies, Regulations and Costs*. 8 December. This report was prepared for the Ministry of Energy and Mineral Resources and included a detailed assessment of 50 geothermal prospects across Indonesia.

[15] This model was prepared by Messrs. Jim Randle and Jim Lawless for the Directorate General of New and Renewable Energy and Energy Conservations under funding from the New Zealand Ministry of Foreign Affairs and Trade, and subsequently updated by the authors for use by PT Sarana Multi Infrastruktur and others.

[16] Presidential Regulation 22/2017.

regarding the cost and potential contribution of geothermal to Indonesia's energy mix. Many of the cheapest geothermal prospects, the "low hanging fruit," have already been developed.

Figure 5:   PLN Implementation of Geothermal Projects Against Plan

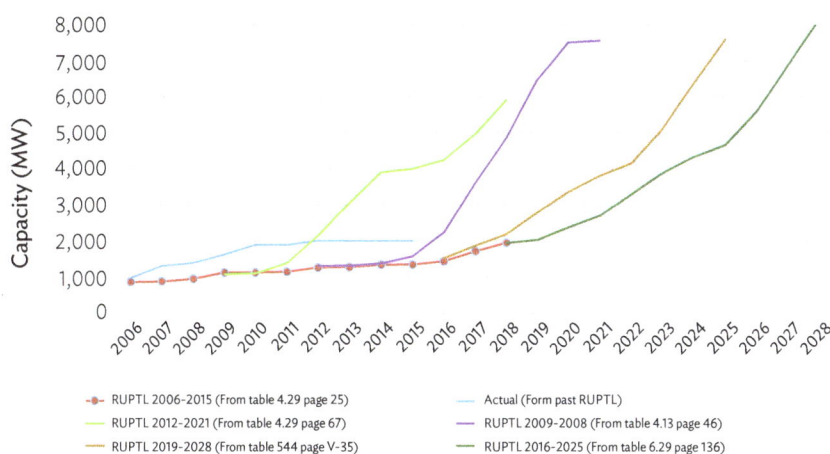

MW = megawatt, PLN = *Perusahaan Listrik Negara* (State Electricity Company),
RUPTL = Rencana Usaha Penyediaan Tenaga Listrik (Electricity Supply Business Plan).

Source: ADB analysis from RUPTL (various years).

The government wishes to minimize subsidies and budget future subsidy obligations with certainty. One way of reducing subsidies is to simply apply a cap on additional subsidies to be awarded in any year or a number of years and allocate this through the reverse auctions. This will reduce the amount of capacity brought online but will allow the lowest-cost projects to proceed. Where auctions are not possible, other mechanisms might be adopted, such as first-come-first-served to encourage the most rapidly deployed projects, or ranking by calculated production cost to select the lowest-cost projects.

These RE subsidies are needed only if the government does not want to pass on to consumers the higher financial costs of greater renewable energy utilization. Alternatively, the government may wish to cap total power sector (tariff plus RE) subsidies at the level of current tariff subsidy, i.e., Rp55 trillion to Rp60 trillion. In that case, some tariff increase would be needed. Assuming the RUPTL 2019–2028 forecast for 2019 sales of 245 terawatt-hour (TWh) along with a total renewable electricity subsidy of Rp6.3 trillion as estimated for 2028, electricity tariffs would only need to increase by Rp26/kWh on average. Such a slight increase (about 2.3% of the current average tariff) could be phased in over time without any measurable adverse economic impacts since the renewable subsidy would not reach Rp6.3 trillion for years to come.

3.11   Summary of the Mechanism

A summary of the recommended budget subsidy mechanism is provided in Table 3.

Table 3:   Summary of Budget Support Mechanism

Component	Summary
Legal basis	• Compensation for a PSO imposed on PLN to procure renewables generation to meet national targets
Governance	• Annual subsidy estimate prepared by PLN • Reviewed and approved by ESDM, MOF, and ultimately DPR for budgeting in the APBN • "True-up" at the end of the year for differences between forecast and actual (audited) RE project prices and volumes
Subsidy calculation	• Calculated individually for each RE project and then summed to obtain the total annual RE subsidy estimate • The subsidy is the difference between the RE project price and PLN's financial cost savings from the RE project (as estimated at the date of PPA signature) multiplied by PLN's purchases • A margin is added to the RE subsidy, consistent with legal requirements and current practice with the tariff subsidy. A substantially lower margin than the prevailing 7% margin applied to the tariff subsidy, e.g., on the order of 0.5%, is recommended, as this lower amount would be consistent with the administrative and financial costs PLN actually incurs to administer the framework. **Cap on PPA price** • If the PPA price results from a competitive tender, then the RE project cost used for subsidy calculations is capped at the economic value, • If there is no competitive tender, then the allowed RE project price is capped at the lower of estimated production cost and economic value, calculated separately for each RE technology and PLN grid. This price serves as the PPA price and is not subject to further negotiation by PLN or the government. • The government can also apply a cap on total subsidies payable to maintain fiscal prudence.
Implementation issues	• Existing ESDM regulations should be amended to allow PLN to sign PPAs at prices above the BPP-based cap, if a subsidy is available • The existing ESDM regulations will also need to be amended to replace the use of BPP with estimated financial cost savings and, ideally, to allow the use of auctions as a procurement mechanism

APBN = *Anggaran Pendapatan dan Belanja Negara* (State Budget), BPP = biaya pokok produksi (production cost), DPR = Dewan Perwakilan Rakyat (House of Representatives), ESDM = *Energi dan Sumber Daya Mineral* (Ministry of Energy and Mineral Resources), MOF = Ministry of Finance, PLN = *Perusahaan Listrik Negara* (State Electricity Company), PPA = power purchase agreement, PSO = public service obligation,
RE = renewable energy.

Source: ADB.

4. Implementation

4.1 Budget Subsidy Implementation

An implementing regulation will be required from MOF covering the calculation, approval, payment, and auditing of the budget subsidy as set out in the preceding section. There is also a potential need for:

- A replacement for the existing ESDM regulations on RE pricing, which permits PLN to pay a higher price than a BPP-referenced price when a subsidy is available, and which allows the use of reverse auctions.

- Alternatively, a new regulation issued by MOF allowing PLN to enter into contracts with RE project owners for environmental benefits or similar services, as a means of paying subsidies to RE projects.

There will also be a need to develop various models and assumptions for use in the calculation of the subsidy:

- Production cost models for major RE technologies to be covered by the subsidy mechanism, to be developed and maintained by MOF.

- An economic benefits calculation model and accompanying financial costs savings model for each RE technology and each PLN grid, to be developed and maintained by PLN but subject to review and approval by MOF and ESDM.

- A list of key assumptions to be used to in the economic benefits calculations, to be issued by MOF.

- A list of key assumptions to be used in the avoided costs component of the economic benefits and in the financial cost savings calculations, to be issued by PLN and to be consistent with those used in the RUTPL.

A decision will be required as to which technologies are covered by the mechanism. Initially, we expect that it would focus on geothermal and hydropower projects, in line with MOF's priorities, with potential subsequent extension to other RE technologies.

4.2 Addressing Other Key Impediments

As identified earlier, closing the price–cost gap is necessary but not sufficient to trigger accelerated RE investments. For some technologies, this gap may be very small or even nonexistent, with other nonprice barriers being the cause of slow-paced development. For this reason, it is desirable for the implementation of a budget subsidy mechanism to be accompanied by other reforms. The key reforms are summarized as follows, mapped against the barriers previously identified in Figure 1.

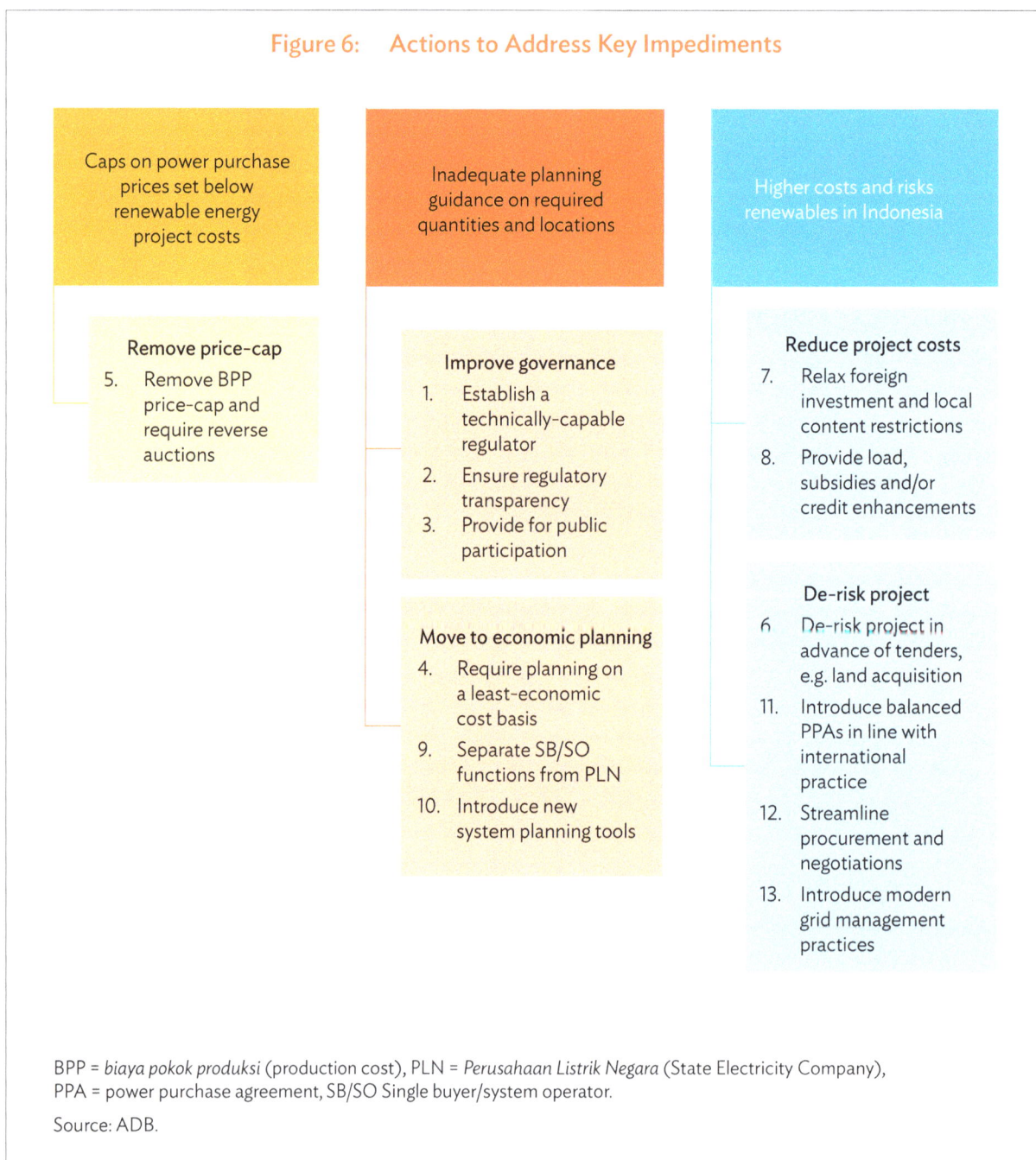

Figure 6: Actions to Address Key Impediments

Caps on power purchase prices set below renewable energy project costs

Remove price-cap
5. Remove BPP price-cap and require reverse auctions

Inadequate planning guidance on required quantities and locations

Improve governance
1. Establish a technically-capable regulator
2. Ensure regulatory transparency
3. Provide for public participation

Move to economic planning
4. Require planning on a least-economic cost basis
9. Separate SB/SO functions from PLN
10. Introduce new system planning tools

Higher costs and risks renewables in Indonesia

Reduce project costs
7. Relax foreign investment and local content restrictions
8. Provide load, subsidies and/or credit enhancements

De-risk project
6. De-risk project in advance of tenders, e.g. land acquisition
11. Introduce balanced PPAs in line with international practice
12. Streamline procurement and negotiations
13. Introduce modern grid management practices

BPP = *biaya pokok produksi* (production cost), PLN = *Perusahaan Listrik Negara* (State Electricity Company), PPA = power purchase agreement, SB/SO Single buyer/system operator.

Source: ADB.

Figure 7 shows which entities would have lead responsibility in each case for implementing these accompanying actions.

Figure 7:   Responsibilities for Addressing Key Impediments

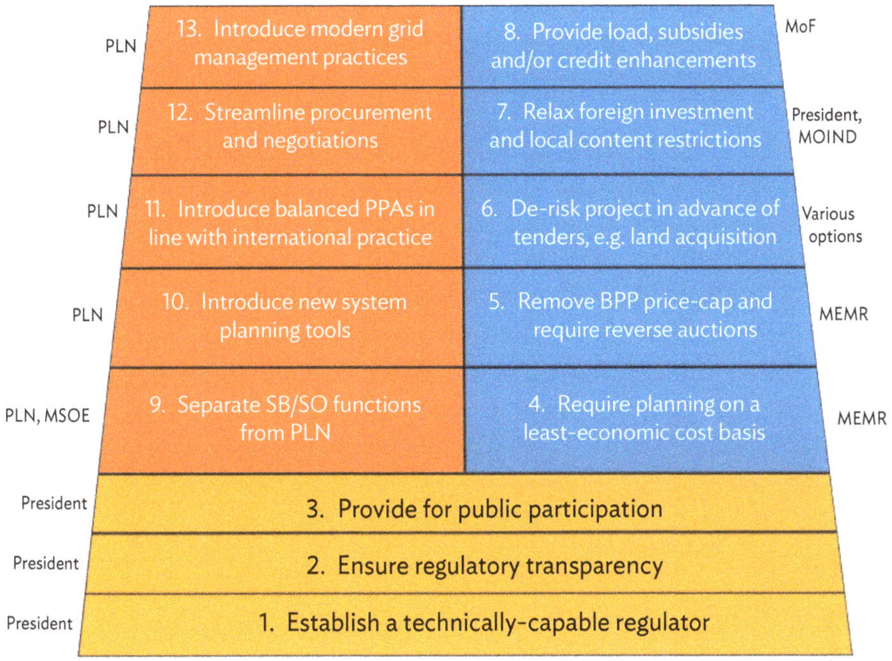

PLN	13. Introduce modern grid management practices	8. Provide load, subsidies and/or credit enhancements	MoF
PLN	12. Streamline procurement and negotiations	7. Relax foreign investment and local content restrictions	President, MOIND
PLN	11. Introduce balanced PPAs in line with international practice	6. De-risk project in advance of tenders, e.g. land acquisition	Various options
PLN	10. Introduce new system planning tools	5. Remove BPP price-cap and require reverse auctions	MEMR
PLN, MSOE	9. Separate SB/SO functions from PLN	4. Require planning on a least-economic cost basis	MEMR

President	3. Provide for public participation
President	2. Ensure regulatory transparency
President	1. Establish a technically-capable regulator

BPP = *biaya pokok produksi* (production cost), MEMR =Ministry of Energy and Mineral Reources, MOF = Ministry of Finance, MOIND = Ministry of Industry,
MSOE = Ministry of State-Owned Enterprises, PLN = *Perusahaan Listrik Negara* (State Electricity Company).

Source: ADB.

Appendix 1:
Indonesia's Renewable Energy Framework

1.1 Policy Objectives and Renewable Energy Targets and Plans

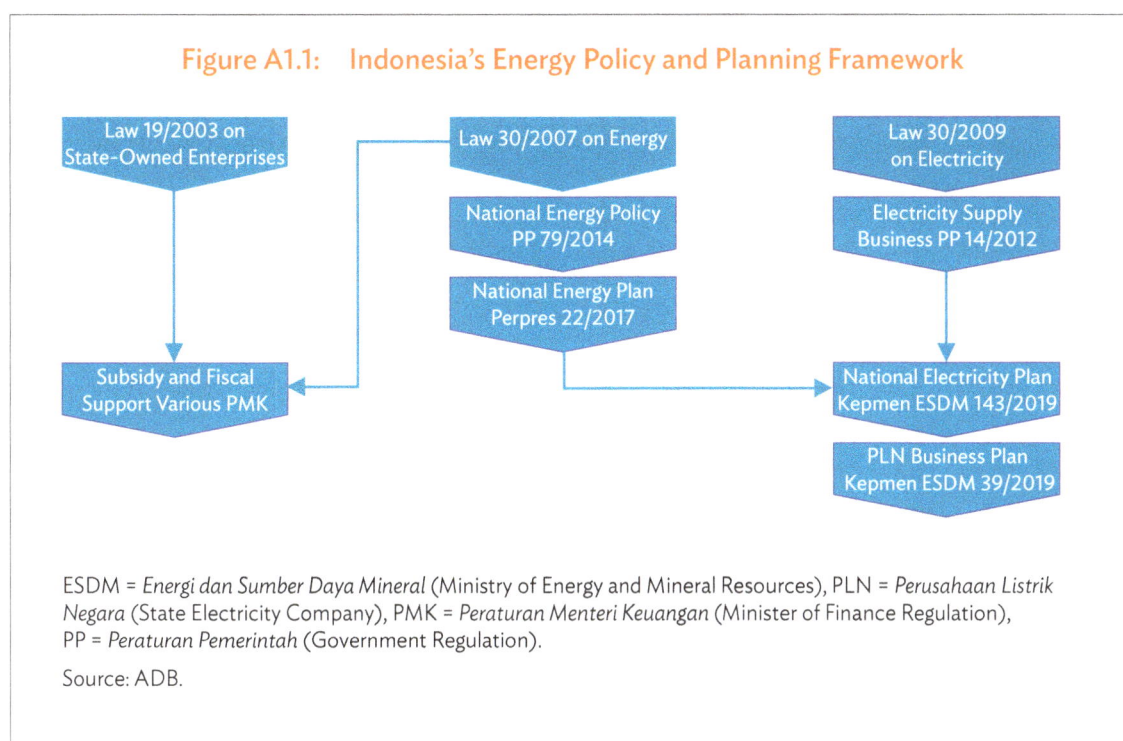

Figure A1.1: Indonesia's Energy Policy and Planning Framework

ESDM = *Energi dan Sumber Daya Mineral* (Ministry of Energy and Mineral Resources), PLN = *Perusahaan Listrik Negara* (State Electricity Company), PMK = *Peraturan Menteri Keuangan* (Minister of Finance Regulation), PP = *Peraturan Pemerintah* (Government Regulation).

Source: ADB.

Law 30/2007 on Energy lays out a cascading framework of energy and electricity policy formulation and planning (Figure A1.1). The figure also shows the linkages with subsidy and fiscal incentive policies as administered by the Ministry of Finance (MOF).

The relevance and of each of these laws and regulations to renewable energy development is as follows:

1. **Law 30/2007 on Energy** establishes the legal basis for the *Kebijakan Energi Nasional* (KEN, National Energy Policy) and the *Rencana Umum Energi Nasional* (RUEN, National Energy Plan). It also provides a basis for facilities and incentives to support renewable energy policy objectives laid out in the KEN and RUEN (Art. 20[5]).

2. **Government Regulation (*Peraturan Pemerintah*, PP) 79/2014** presents the **KEN**. It targets that 23% of total primary energy supply is to be supplied by renewable energy by 2025.

3. **Presidential Regulation (*Peraturan Presiden*, PERPRES) 22/2017** lays out the **RUEN**, which represents a high-level plan to achieve the KEN. It calls for 45.2 gigawatts (GW) of renewable energy capacity to be installed by 2025 to meet the 23% renewable share of total primary energy supply stipulated in the KEN.

4. **Law 30/2009 on Electricity** stipulates that the *Rencana Umum Ketenagalistrikan Nasional* (RUKN, National Electricity Plan) shall be based on the RUEN.

5. **PP 14/2012 on Electricity Supply Business Activities** stipulates that electricity supply licensees involved in retailing, distribution, or vertically-integrated supply must prepare a *Rencana Usaha Penyediaan Tenaga Listrik* (RUPTL, Electricity Supply Business Plan) in accordance with the RUEN.

6. **Minister of Energy and Mineral Resources Decree (*Keputusan Menteri Energi dan Sumber Daya Mineral*, KEPMEN ESDM) 143/2019** presents the RUKN for the period 2019–2038. It includes two scenarios for the target 2025 fuel mix, one scenario as business-as-usual (BAU), the other with more aggressive conservation efforts. In the BAU scenario, 76 GW of generating capacity are to be added by 2025, of which 22 GW should be renewable. In the conservation scenario, 67 GW of new capacity are to be added, of which 17 GW should be renewable. These additions are on top of the 6.7 GW already installed by the end of 2018.

7. **KEPMEN ESDM 39/2019 on Endorsement of State Electric Company (*Perusahaan Listrik Negara*, PLN) RUPTL 2019–2028** was issued prior to the release of KEPMEN ESDM 143/2019, and therefore refers to the RUKN 2008–2027. It adopts a lower demand forecast than the RUKN and projects 48.6 GW of capacity additions by 2025, of which 14.3 GW renewable capacity is planned.

8. **Law 19/2003 on State-Owned Enterprises (SOEs)** stipulates that the government may appoint a state-owned enterprise to conduct activities for the public good. Because SOEs are established to seek a profit, if this appointment results in an obligation that is not financially feasible for the SOE, then the government is required to compensate the SOE for the expenses it incurs to execute the appointment plus a margin.

9. The Minister of Finance issues various regulations (*Peraturan Menteri Keuangan, PMK*) regarding the administration of compensation paid to SOEs with respect to financial infeasible public services as contemplated under Law 19/2003, as well as fiscal incentives for renewable energy development as contemplated under Law 30/2007.

Further to this domestic energy policy agenda, Indonesia has ratified the Paris Agreement under **Law 16/2016 on Ratification of the Paris Agreement to the United Nations Framework Convention on Climate Change**, in which Indonesia commits to a Nationally Determined Contribution of 29% reduction in greenhouse gas emissions on its own by 2030, or 41% reduction with international cooperation.

Table A1.1 compares the existing generation capacity with the amounts targeted/planned under each of the principal plans discussed. The decreasing renewable capacity target is consistent with the reduction in total generation capacity requirements resulting from slower load growth. The RUEN, for example, is based on unrealistically high economic growth rates and income elasticities. In preparing the RUPTL 2019–2028, PLN has utilized a more realistic demand forecast, which reduces total capacity requirements as well as planned renewable capacity.

Table A1.1:   Renewable Penetration Targets and Load Forecasts

Source	Targeted/Planned 2025 Renewables (GW)	Total Targeted/Planned 2025 Capacity (GW)	Renewable (%)
End-2018 actual generation mix	6.7	51.2	13.1%
RUEN	45.2	135.5	33.4%
RUKN 2019-2038 (BAU)	28.7	122.0	23.5%
RUPTL 2019-2028	21.0	99.8	21.0%

BAU = business-as-usual, RUEN = *Rencana Umum Energi Nasional* (National Energy Plan), RUKN = *Rencana Umum Ketenagalistrikan Nasional* (National Electricity Plan), RUPTL = *Rencana Usaha Penyediaan Tenaga Listrik* (Electricity Supply Business Plan).

Note: Targeted/Planned values are based on de-rated capacity (*daya mampu netto*).

Source: RUEN, RUKN 2019-38, and RUPTL 2019-28.

Figure A1.2:   Renewable Power Capacity in Selected Countries

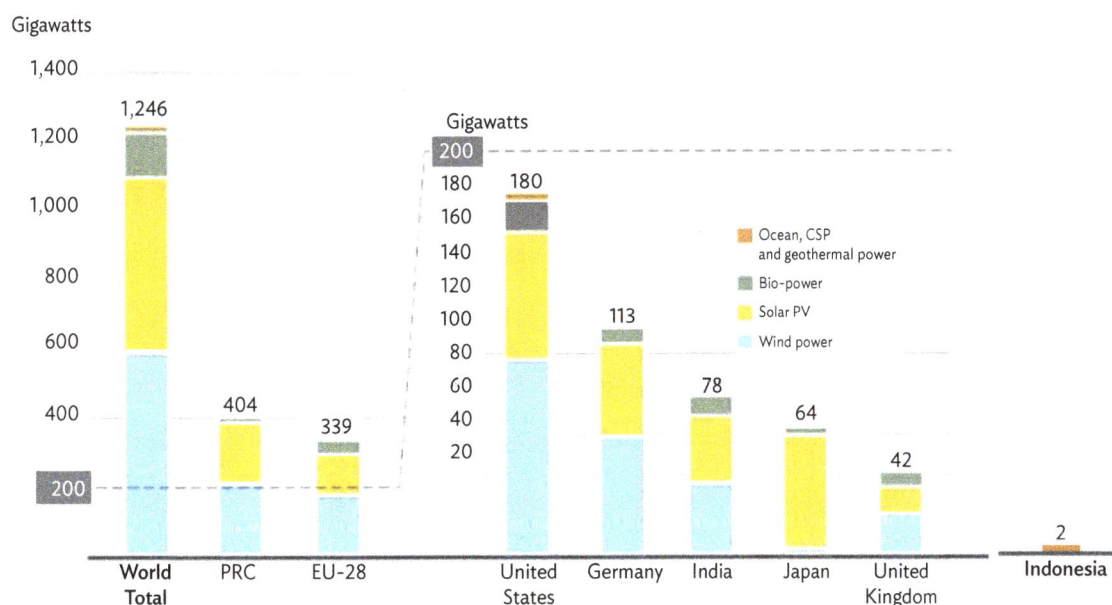

Note: Graph does not include hydropower. *The EU-28 refers to Austria, Belgium, Bulgaria, Croatia, Cyprus, the Czech Republic, Denmark, Estonia, Finland, France, Germany, Greece, Hungary, Ireland, Italy, Latvia, Lithuania, Luxembourg, Malta, the Netherlands, Poland, Portugal, Romania, Slovakia, Slovenia, Spain, Sweden, and the United Kingdom.* CSP = concentrated solar power, PRC = People's Republic of China, PV = photovoltaic.

Ren21, *Renewable 2019 Global Status Report*

1.2   Performance Against Targets

Indonesia has adopted aggressive renewable energy development plans but, compared to other countries, actual uptake has been slow. Figure A1.2 compares Indonesia's installed renewable capacity (excluding hydro) against countries that have aggressively developed renewable energy.

Figure A1.3 compares variable renewable energy penetration in selected countries.

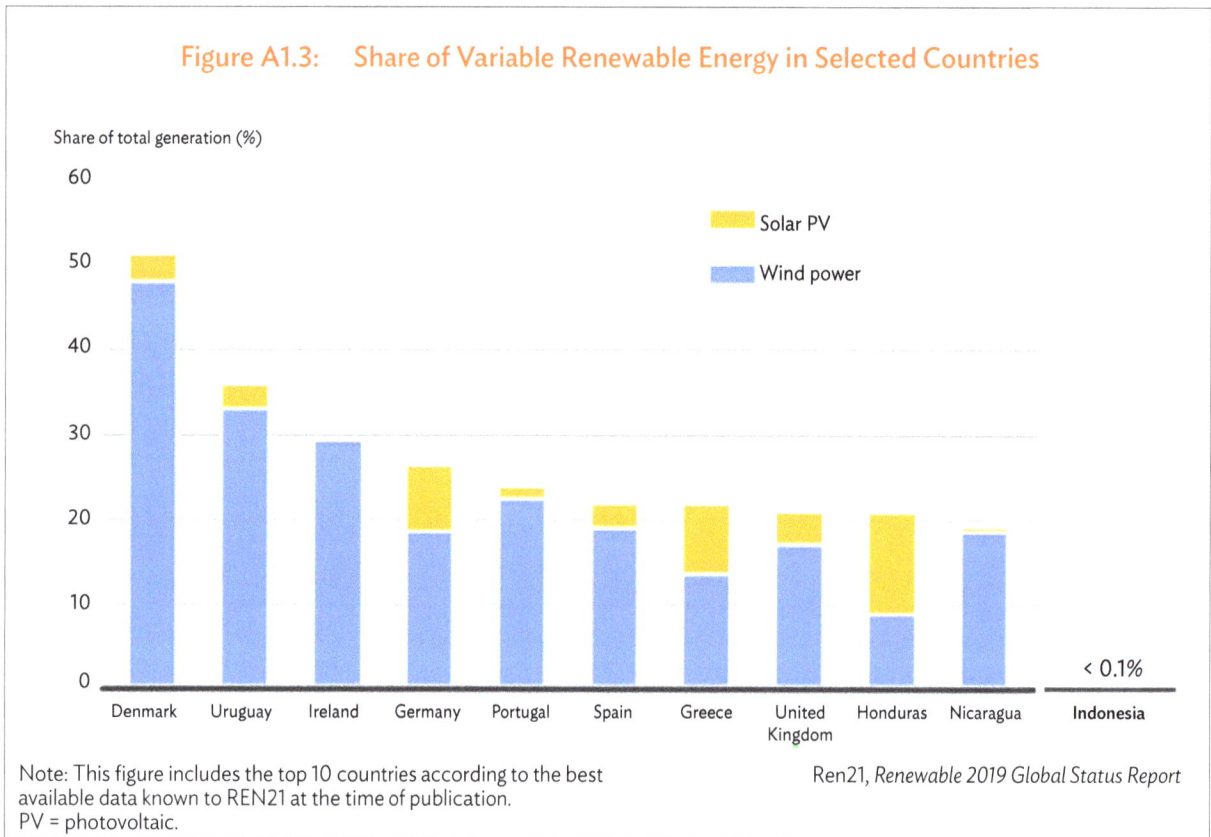

Figure A1.3:   Share of Variable Renewable Energy in Selected Countries

Note: This figure includes the top 10 countries according to the best available data known to REN21 at the time of publication.
PV = photovoltaic.

Ren21, *Renewable 2019 Global Status Report*

Geothermal and hydro are the dominant grid-connected renewable energy sources at present and as planned for the future in Indonesia. By the end of 2018, geothermal and hydro represented 96.2% of all grid-connected renewable power generation based on nameplate capacity. The remaining 3.8% (264 MW) was supplied by wind, biomass, and solar. The RUPTL 2019–2028 maintains this dominant role and aims for geothermal and hydro to provide 85.7% renewable capacity by the end of 2025.

Over the past 5 years, Indonesia has added an annual average of 171 MW of hydro capacity and 134 MW of geothermal capacity. To achieve the end- 2025 targets enumerated in the RUPTL 2019–2028 for these technologies, Indonesia would need to add on average 908 MW of hydro capacity and 727 MW of geothermal capacity per year, more than five times the annual rate of capacity additions for these technologies over the previous 5 years.

Figure A1.4: Actual and Planned Hydro Capacity, 2006–2028

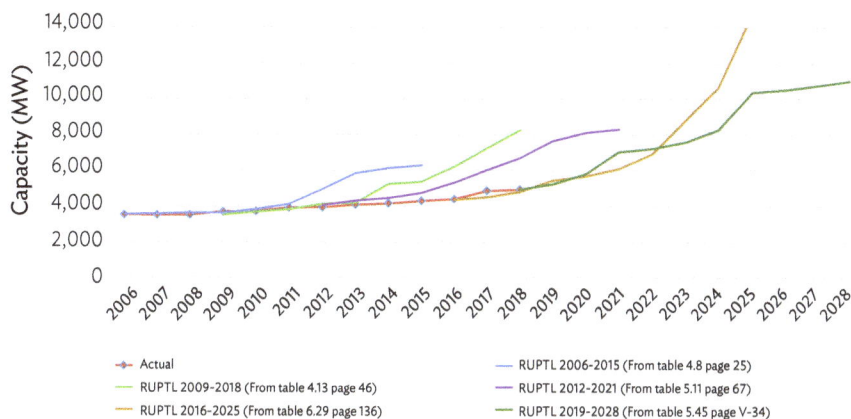

Legend:
- Actual
- RUPTL 2009-2018 (From table 4.13 page 46)
- RUPTL 2016-2025 (From table 6.29 page 136)
- RUPTL 2006-2015 (From table 4.8 page 25)
- RUPTL 2012-2021 (From table 5.11 page 67)
- RUPTL 2019-2028 (From table 5.45 page V-34)

MW = megawatt, RUPTL = *Rencana Usaha Penyediaan Tenaga Listrik* (Electricity Supply Business Plan)
Source: ADB analysis from RUPTL (various years).

Figure A1.5: Actual and Planned Geothermal Capacity, 2006–2028

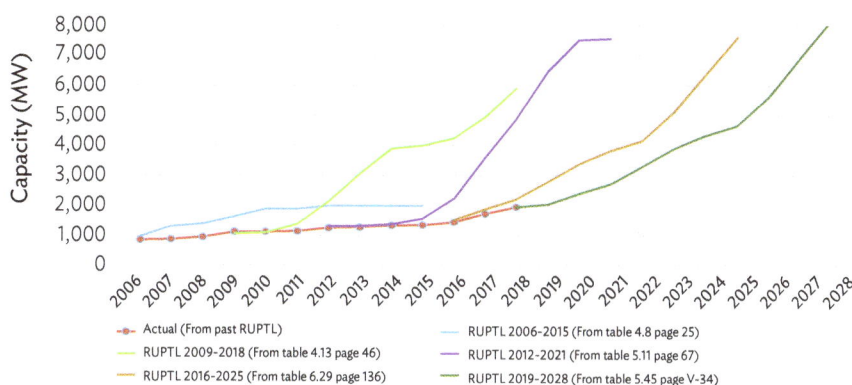

Legend:
- Actual (From past RUPTL)
- RUPTL 2009-2018 (From table 4.13 page 46)
- RUPTL 2016-2025 (From table 6.29 page 136)
- RUPTL 2006-2015 (From table 4.8 page 25)
- RUPTL 2012-2021 (From table 5.11 page 67)
- RUPTL 2019-2028 (From table 5.45 page V-34)

MW = megawatt, RUPTL = *Rencana Usaha Penyediaan Tenaga Listrik* (Electricity Supply Business Plan)
Source: ADB analysis from RUPTL (various years).

As shown in Figures A1.3 and A1.4, PLN has planned substantial renewable capacity additions over the past 15 years but has never achieved these targets over the longer term. Clearly, there is a disconnect between the country's ambitious plans to scale up renewable energy supply and its laggard performance.

1.3 Enabling Conditions for Renewable Energy Development

Successful exploitation of renewable energy depends on six factors as shown in Figure A1.6.

Most of these factors are sufficiently present in Indonesia to create a favorable environment for renewable energy development. But as discussed here, the principal impediment to greater uptake of renewable energy lies in government policies and regulations.

(i) **Resource base.** This refers to the physical availability of renewable energy resources in the country. Indonesia is blessed with a wide range of renewable energy resources. Given its tropical location, insolation is not as good as in desert regions of the world, but with annual average insolation of 4.5 to 5.5 kilowatt-hours per square meter per day (kWh/m2/day) for about two-thirds of the nation's land mass,[17] solar power can be developed commercially. Much of the country is mountainous and receives adequate rainfall for hydro development. While the commercial potential for geothermal is certainly lower than the 17,506 MW of reserves and 11,073 MW of resources cited by ESDM,[18] Indonesia hosts the world's largest collection of high-enthalpy hydrothermal prospects in the world. There are also commercially viable wind resources in certain parts of the country, with potential estimated to be as much as 9.3 GW.[19] The fertile soil supports biomass production, and as an archipelagic nation there is also potential for ocean power.

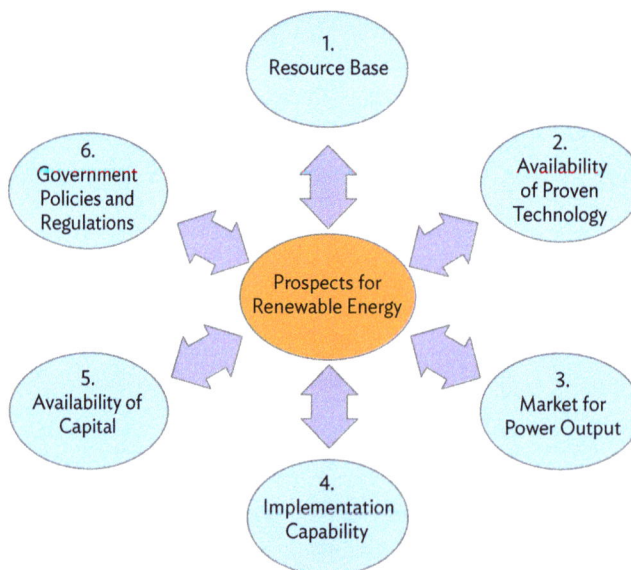

Figure A1.6: Uptake of Renewable Energy Depends on Six Factors

Source: ADB.

---

[17] World Bank Group (ESMAP), *Solar Resource and Photovoltaic Power Potential of Indonesia,* May 2017,

[18] Tempo. 2018. *Indonesia only utilizes 10 percent of geothermal energy.* 14 December. https://en.tempo.co/read/908346/indonesia-only-utilizes-10-percent-of-geothermal-energy.

[19] International Renewable Energy Agency. 2017., *Renewable Energy Prospects: Indonesia, a REmap analysis.* www.irena.org/remap.

(ii) **Availability of proven technology.** Conversion of naturally occurring renewable resources into electricity requires proven technologies. Commercially proven generation technology is available from international markets for most of the renewable energy resources that are widely found in Indonesia.

(iii) **Market for p**ower output. In 2018, PLN power sales reached 235 terawatt-hours (TWh). RUPTL 2019–2028 forecasts power sales to grow at an annual average of 6.4% over the coming 10-year planning period. As shown in Figure A1.7, PLN has already committed to sufficient generation additions to meet total PLN demand as forecast through 2024, assuming a national average reserve margin target of 30%. This includes 3,962 MW of committed renewable generation capacity additions from 2019 through 2024, resulting in a 15% renewable share of total derated PLN capacity in 2024. PLN's plan to reach installed renewable capacity of 12,434 MW by end of 2028 (a 21% share of total derated generation capacity) depends on planned (but not yet committed) renewable additions of 7,519 MW. However, as highlighted in the previous chapter, PLN has demonstrated a chronic inability to achieve its generation expansion plans. Realization of these planned but not yet committed additions will depend on a supportive enabling regulatory and policy environment.

(iv) **Implementation capability**. Development and implementation of renewable energy projects requires human resources, land, and logistical networks. Indonesia has many qualified engineering, procurement, and construction contractors. Some local developers for small projects (less than 10 MW) may lack sufficient understanding of project risks and development requirements, but this could be addressed by allowing higher levels of foreign investment in small projects to attract more sophisticated international developers to partner with local firms. Land access has improved in recent years, but can still be challenging; it is not uncommon for infrastructure projects to face delays due to the difficulty of securing access, and in some cases the projects are unable to proceed. [20] In most cases, however, land access can be secured albeit with additional time, effort, and cost. Logistics in Indonesia remains a work in progress, but it is adequate to support development of renewable energy resources throughout the country.

Figure A1.7: PLN Sales Forecasts and Committed and Planned Capacity Additions

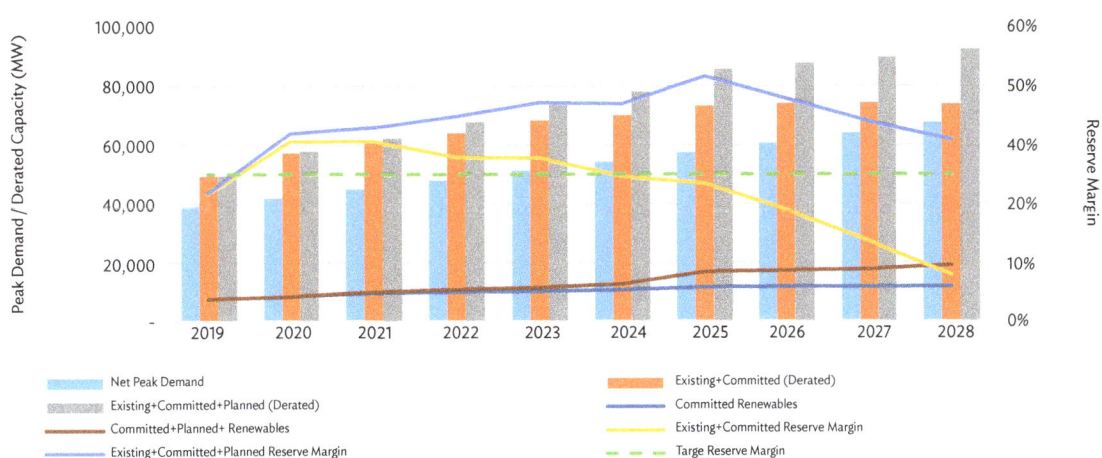

MW = megawatt, PLN = *Perusahaan Listrik Negara* (State Electricity Company).
Source: ADB analysis from RUPTL 2019–2028.

---

[20] For example, see M. Iswara. 2019. Land 'syndicates' blamed for driving investors away from RI. *Jakarta Post*. 15 October. https://www.thejakartapost.com/news/2019/10/15/land-syndicates-blamed-driving-investors-away-ri.html.

(iv)   **Availability of capital.** There is a wide range of commercial banks, multilateral and bilateral finance institutions, domestic nonbank financial institutions like Sarana Multi Infrastruktur (SMI), and private investors who are prepared to lend, and some cases take equity positions in projects. Available funding for renewable energy projects has not been fully utilized.

(v)   **Government policies and regulations.** Renewable energy development is subject to a wide range of government policies and regulations spanning agencies. Unfortunately, interministerial coordination is weak and there is no apparent analysis of policy and regulatory impacts prior to the implementation of new policies and regulations. Consequently, policies and regulations are often issued to address one immediate concern without due consideration of longer-term impacts or the need for a unified strategy across ministries.

The adequacy of the preceding factors is, in fact, negated by counterproductive government policies and regulations. At least, this means that creating a more conducive renewable energy development environment is within the control of the government.

The principal agencies involved in renewable energy policy making and regulation are as follows:

- MOF administers subsidies and fiscal incentives;

- ESDM handles economic and technical regulation of the power sector, including licensing, planning, and tariff setting both for retail and bulk power supply;

- Ministry of Industry sets local content requirements;

- Ministry of State-Owned Enterprises appoints and supervises the management of state-owned enterprises including the national utility, PLN; and

- the President issues the negative investment list and holds authority that supersedes each of the ministerial authorities enumerated above.

Current conditions with respect to each of these policy and regulatory areas are described in further detail as follows.

1.4   Subsidies and Fiscal Incentives

MOF is responsible for the administration of subsidies, fiscal incentives, and other direct government support. The legal basis for these forms of support is provided by:

- **Subsidies.** Law 19/2003 on State-Owned Enterprises (SOEs), which obliges the government to compensate an SOE for the execution of any government assignment for the public good that is not profitable. This compensation consists of unrecovered costs plus margin. Although there are no renewable energy subsidies at present, examples of other energy subsidies administered by MOF under this law include:

  - Minister of Finance Regulation (PMK) 116/2016 on Procedures for the Budgeting, Calculation, Payment and Responsibility for Funds for the Liquified Petroleum Gas Subsidy; and

  - PMK 44/2017 on Procedures for the Budgeting, Calculation, Payment and Responsibility for Funds for the Electricity Subsidy.

- **Tax incentives.** Law 30/2007 on Energy authorizes the government to provide facilities and incentives to companies and individuals for renewable energy supply. This follows various laws on taxes and duties, which allow for tax facilities for strategic activities with respect to income tax, value-added tax (VAT), and import taxes and duties. The specific implementing regulations are as follows:

  o **Tax and duty allowances.** There is a suite of government regulations and Minister of Finance regulations[21] that provide tax and duty allowances for greenfield investment in renewable energy:[22]

    - an investment tax deduction equivalent to 30% of fixed capital investment, applied as 5% over 6 years;

    - accelerated depreciation and amortization;

    - exemption from Article 22 import tax on machines and equipment, excluding spare parts; depending on the imported good, this can be as much as 7.5% of the declared value;

    - VAT exemption on imported goods, excluding spare parts;

    - import duty exemption;

    - reduction of tax on dividends remitted to non-residents to 10% or less depending on the prevailing tax treaty; and

    - extension of tax loss carry forward from 5 years up to 10 years, subject to certain criteria.

  o **Tax holiday.** PMK 150/2018 on Provision of Corporate Income Tax Reduction Facilities provides corporate income tax holidays for investment in "pioneer industries" including "economic infrastructure," which includes renewable energy power plants as stipulated in Badan Koordinasi Penanaman Modal (Investment Coordinating Board) Regulation 1/2019 as amended by 6/2019.

- **Direct support.** There are ways the government can directly fund renewable energy development.

  o **Line ministry and regional infrastructure budgets.** This mechanism is not disbursed by MOF but by line ministries like ESDM or the Ministry of Villages, Disadvantaged Regions, and Transmigration, which may directly fund the implementation of off-grid renewable energy projects. This funding comes from the State Budget (APBN), which is compiled by MOF with inputs from line ministries and ultimately approved by the national legislature (DPR). The 2019 APBN budgets approximately Rp1 trillion for ESDM's Directorate General of New and Renewable Energy and Energy Conservation (*Energi Baru dan Terbarukan dan Konservasi Energy,* EBTKE) to plan, implement, and supervise renewable energy projects, which typically are not grid connected, e.g., solar home systems or community PV systems. Funding can also be disbursed through the Special Allocation Fund (*Dana Alokasi Khusus,* DAK), which provides funding to regional governments from the APBN for specific activities, like off-grid renewable electricity supply.

---

[21] These include: Government Regulation (*Peraturan Pemerintah, PP*) 18/2015 as amended by 9/2016 on Income Tax Facilities for Investment in Certain Industries and/or Regions; PMK 89/2015 as amended by on Procedures for Provision of Income Tax Facilities for Investment in Certain Industries and/or Regions as well as Transfer of Assets and Sanctions for Domestic Taxpayers; PMK 21/2010 on Provision of Tax and Duty Facilities for Renewable Energy Activities; PMK 176/2009 as amended by PMK 188/2015 and PMK 76/2012 on Import Duty Exemption for Investment in Equipment, Goods and Materials for Industrial Development.

[22] "Renewable energy" specifically includes geothermal, wind, biomass, solar, hydro and ocean energy.

- o **Public–private partnerships.** MOF manages direct government support to public–private partnership (PPP) projects. However, given the minimum project sizes required for this support and the additional requirements to qualify as a PPP project (including the sponsorship and partnership of a national or regional government agency), PPP arrangements are not well suited for renewable energy development. The specific mechanisms include:

  - a project development facility that can fund PPP project preparation; and

  - viability gap financing, which can fund up to 49% of construction costs for qualifying PPP projects with a minimum investment of Rp100 billion.

- **Government guarantees,** which may be issued directly by MOF as a Business Viability Guarantee Letter, or by the Indonesian Infrastructure Guarantee Fund (IIGF), an SOE under MOF. In addition to the availability of IIGF guarantees for PPP infrastructure projects, this mechanism has also been used to secure financing for SOEs from international development finance institutions that require sovereign or quasi-sovereign guarantees as a basis for infrastructure lending. A recent example is the guarantee extended by the IIGF for KfW's hydropower loan to PLN.

- **Environmental Fund Management Agency (*Badan Pengelola Dana Lingkungan Hidup,* BPDLH).** BPLDH was established based on Government Regulation 46/2017 on Economic Instruments for the Environment, and PERPRES 77/2018 on Environmental Fund Management. It has been established as a public service agency (*badan layanan umum,* BLU), so that it can manage future receipts and expenditures for environmental management activities outside of the APBN. It will be supervised by an interministerial steering committee, chaired by the Coordinating Minister of Economy, including MOF and ESDM as members, with the Ministry of Forestry and Environment serving as secretariat.

  BPDLH is intended to finance and manage climate change activities in accordance with Indonesia's commitments under the Paris Agreement, which was ratified as Law 16/2016. Potential future funding sources include environmental taxes and levies, such as a carbon tax. Expenditure is authorized for carbon trading, loans, grants, and subsidies, and must be based on an agreement or contract between BPDLH and the beneficiary. BPDLH will start operation on 1 January 2020 with initial funding of Rp2.1 trillion obtained through merger with the Ministry of Forestry and Environment's Forestry Financing and Development BLU.

- **Geothermal resource risk mitigation facility.** A geothermal resource risk mitigation facility of $455 million has been established with funding from the Government's Infrastructure Financing Facility for the Geothermal Sector (*Pembiayaan Infrastruktur Sektor Panas Bumi,* PISP), the World Bank, the Clean Technology Fund, the Green Climate Fund, and private developer equity. This facility is managed by SMI and is available to both private and state-owned developers, though the allocation of funding and activities eligible for support (exploration and delineation drilling) between private and public sector development efforts has not been finalized.

  SMI is managing this resource risk mitigation facility together with the Geothermal Energy Upstream Development Program (GEUDP), which funds exploration drilling for smaller, unproven geothermal prospects. These results are then provided to ESDM as a basis for tendering out the prospect. The first prospect supported by GEUDP is Waesano in Flores, Nusa Tenggara Timur.

1.5   Tariffs, Procurement, and Contracting

ESDM is responsible for power sector policy making and regulation. This includes regulation of tariffs, as well as procurement and contracting of renewable energy generation by PLN.

ESDM's overriding power sector policy objective since the start of 2017 has been to ensure affordability of electricity. As a result, ESDM suspended application of the automatic tariff adjustment mechanism, which was most recently stipulated in PERMEN ESDM 41/2017, and retail electricity tariffs have not been adjusted since then.

As part of this effort to ensure affordability, ESDM simultaneously implemented regulations on the supply side, capping the price of most renewable technologies at some percentage of PLN's regional generation production cost (BPP). Other ESDM regulations are presumably intended to enhance government control over renewable energy development, but appear to have resulted in making renewable energy development by private parties more expensive or difficult. Principal ESDM regulations governing renewable energy development are described in further detail as follows.

- **PERMEN ESDM 50/2017 on Utilization of Renewable Energy Resources for Electricity Supply**, as amended by 53/2018,[23] governs the pricing and procurement of renewable energy sold to PLN. Key aspects of this regulation are discussed as follows:

  - **Price formation.** Renewable energy power prices are negotiated between PLN and the developer, and subsequently approved by ESDM (which may involve renegotiation). If the project is located in an area where the regional BPP is greater than national average BPP, then the negotiated price cannot exceed either 85% or 100% of the regional BPP depending on the technology. BPP is calculated by PLN region (roughly corresponding to provinces) based on PLN's accounting cost of generation for the previous year. BPP does not represent the economic value of renewable power generation to Indonesia for the following reasons:

    - BPP is based on historical costs from the previous year (backward looking), whereas the value of any new generation should take into account the expected cost of fuel in the future (forward looking);

    - BPP uses depreciation, which is typically lower than actual principal repayments that may be required to finance new generation and based on the value of old plants that can be replaced at their original book value.

    - As a pure accounting cost, BPP ignores economic value, externalities, and associated social costs. For example, it uses the price PLN pays for fuel. Most PLN generation is from coal, and ESDM has capped coal prices to PLN at a benchmark of $70/ton. In 2018, international cost of benchmark coal reached over $100/ton. Moreover, there are several negative externalities associated with coal-fired power generation that are not reflected in the price PLN pays, but which should be taken into account when evaluating options for new generating capacity.

    - BPP represents only the direct costs of generation (e.g. depreciation, financing costs, fuel, and operation and maintenance) and does not include overhead expenses or profit margin.

    - BPP is calculated as an average cost across all generation in the region, but pricing of new generation should be based on the marginal costs of alternatives.

---

[23]   PERMEN 53/2018 only adds power generation using liquid biofuel to the scope of PERMEN 50/2017.

- o   **Procurement modality.** Other than waste-to-energy and geothermal,[24] projects are procured through direct selection, i.e., PLN compares bids from prequalified developers.

- o   **Form of investment.** Until 2020, all renewable energy projects that sell power to PLN (with the exception of waste-to-energy) had to be structured as a build–own–operate–transfer (BOOT), so that developers had to transfer the project to PLN at the end of their power purchase agreement (PPA). This prevented developers from recognizing any residual value in their projects, resulting in the need to secure higher prices for power produced by the project during the term of its PPA.

- o   **Regulatory certainty.** PERMEN 50/2017 represents the most recent of a series of shifting regulatory conditions. For example, ESDM has issued eight geothermal pricing regulations over 10 years: PERMEN ESDM 14/2008, 5/2009, 32/2009, 2/2011, 22/2012, 17/2014, 12/2017, and 50/2018. Given that geothermal projects typically require 5 to 7 years from initial exploration (and the associated investment) to commissioning, the continually changing regulatory environment increases risks for developers, and may well cause some investors to avoid Indonesia.

- •   **PERMEN ESDM 10/2017 on Main Points of Power Purchase Agreements (PPAs),** as amended by 49/2017 and 10/2018, prescribes the contents of PPAs for dispatchable power projects, including geothermal, biomass, and hydro. It does not apply to wind, solar, hydro less than 10 MW, biogas, and waste-to-energy (though it appears that PLN has applied these same conditions to the negotiation of PPAs using these technologies). The following are key aspects of the regulation as most recently amended:

  - o   Although the latest amendment removes references to government force majeure (GFM), which presumably allows PLN and the IPP to negotiate how GFM will be handled under the PPA, PLN is relieved of obligations in the event of natural force majeure that disrupts the ability of the grid to absorb the power generated by the IPP. Previously, IPPs would receive deemed dispatch payments in the event they were available but could not evacuate power due to problems with the grid.

  - o   Payments must be made to IPPs in Indonesian rupiah. PLN interprets this to mean that it cannot guarantee the convertibility of the payments into foreign currency, and that its payment obligation is fulfilled once the rupiah payment is made, and places the risk of US dollar availability and conversion onto the developers and lenders, which pushes up the debt and equity costs.

  - o   BOOT requirement. This regulation requires all covered generation to be developed on a BOOT basis. As noted above, this increases the cost of power supply from IPPs. However, the BOOT requirement was eliminated under PERMEN 4/2020.

- •   **PERMEN ESDM 37/2018 on Offering of Geothermal Working Areas, Issuance of Geothermal Permits, and Appointment of Geothermal Businesses.** This regulation is specific to geothermal power development. One of the key elements of this regulation is that PPAs are now signed only after exploration is complete, a commercial resource is proven, and a feasibility study is prepared. In practice, this means that geothermal developers must invest tens of millions of dollars in exploration before knowing what price they will receive for power in the event a commercially viable resource is proven. This uncertainty about PPA terms and conditions and the price that would be paid creates additional risks for developers, which consequently require higher returns and therefore higher prices—or deter developers

---

[24]   Waste-to-energy and geothermal are procured through direct appointment, i.e., sole source procurement. However, new geothermal prospects opened to private developers are awarded based on competitive tendering so that the private developer that finds a commercially viable prospect has already been determined through a competitive process.

entirely from pursuing projects. Previously, PPAs could be signed after the completion of the preliminary survey but prior to the commencement of costly exploration drilling.

1.6   Local Content Requirements and Investment Restrictions

President Regulation (*Peraturan Presiden*, PERPRES) 44/2016 on List of Business Areas Closed to Capital Investment and Areas Open with Conditions, otherwise known as the "negative investment list," stipulates the maximum foreign investment in various types of projects. For power generation projects (including renewable energy):

- less than 1 MW no foreign investment is allowed;

- 1 to 10 MW, foreign ownership may not exceed 49%;

- more than 10 MW, foreign ownership may not exceed 95%, unless it is a public–private partnership, in which foreign ownership up to 100% is allowed; and

- an exception is made for geothermal power projects equal to or less than 10 MW, foreign ownership may not exceed 67%.

Minister of Industry Regulation (PERMEN Perin) 54/2012 as amended by 5/2017 on Use of Domestic Product for Development of Electricity Infrastructure, as amended by PERMEN Perin 5/2017 stipulates minimum domestic content requirements for geothermal, hydro, and solar PV projects that will generate power for public consumption, and are funded by the state budget, regional government budget, grant, or foreign loan. This requirement applies to projects carried out by state-owned enterprises, regional government-owned enterprises, cooperatives, and private companies.

However, as acknowledged by ESDM, domestic industry is not yet capable to meet these targets.[25] In 2018, for example, local content for

- hydro projects between 15 and 150 MW achieved only 33.78% local content compared to a target of 59.88%,

- solar PV projects achieved only 35.25% versus a target of 42.2%, and

- geothermal projects from 5 to 110 MW achieved only 17.29% versus a target of 26%.

PERPRES 24/2018 on the National Team for the Increased Utilization of Domestic Products established a ministerial-level team (Tim Nasional Peningkatan Penggunaan Produk Dalam Negeri, P3DN) to enforce domestic content requirements.

However, the application of domestic content regulations for renewable energy faces a chicken-and-egg conundrum. Investment for domestic production of renewable energy equipment will come if the market is large enough. But, the market growth is constrained by the high cost of locally produced equipment, which has yet to achieve economies of scale. The success in applying local content regulations for mobile phone production

25   F. Sulmaihati. 2019. Selama 2018, Hanya PLTU yang Penuhi Target Penggunaan Produk Lokal". *Katadata*. 11 January. https://katadata.co.id/berita/2019/01/11/selama-2018-hanya-pltu-yang-penuhi-target-penggunaan-produk-lokal.

in Indonesia has been a success because the market had been allowed to develop to a point where domestic production could achieve the economies of scale required to keep prices affordable.

Moreover, the limitation of foreign investment to a minority interest in projects 1 to 10 MW inhibits participation of more sophisticated foreign developers who presumably have access to additional financing sources and understand better how to mitigate project risks. As a result, one of the main challenges for development of projects in the 1 to 10 MW range is not financing, but the proper preparation of projects that can attract financing.

PLN's current practices raise other issues as well:

- A shift in the recent Bali solar tenders toward PLN only agreeing to a portion of the capital cost component being indexed to US dollar (approximately 60%), with the balance being Indonesian rupiah-only. Much like the local content market, the local bank/project bond market is not there yet in terms of providing cost competitive long tenor fixed-rate loans.

- PLN is constantly coming out with new "generations" of PPAs which adds transaction costs and slows down the procurement process. Standardized PPAs consistent with international practice and allocation of risks to the parties best able to manage them are long overdue. Other Asian countries can implement very fast-track procurement processes using "non-negotiable" standard PPAs, but the key is to ensure that the standard is bankable. PLN has many examples of bankable PPAs that have been successfully financed (therefore acceptable to PLN, developers, and lenders), so the development of a standard template should not be a burdensome exercise.

- PLN has in recent years made it increasingly difficult for sponsors to sell down their interests in the projects (no sale without PLN approval, only sales to someone who is a registered developer on PLN's List of Selected Suppliers, or PLN insisting on a right to match). This increases the required returns and project costs by preventing developers from recycling capital after commercial operations date is achieved by, for example, selling to a pension fund interested in stable long-run returns at lower rates.

Appendix 2:
Illustrative Subsidy Calculations

2.1 Distinguishing Between Economic Values and Financial Costs

A constant theme throughout this chapter and in subsequent chapters discussing the need for and level of subsidies for power purchases from renewables is the distinction between economic values and financial costs. Put simply, the two can be defined as follows:

- **Economic values** represent the costs and benefits to society, rather than to the project's owners who are just one part of society. A project should only proceed if these social benefits exceed the social costs.

- **Financial costs** represent the costs and revenues to the project's owners. Project owners will only proceed with a project if its financial revenues exceed its financial costs.

It follows from this that there are significant differences between the calculation of economic values and financial costs. Some of the most important of these are briefly discussed as follows:

- **Inclusion of externalities.** The most obvious difference is that an economic valuation includes costs and benefits associated with project impacts that are not reflected in its financial costs or revenues "externalities". A typical example of such costs would be pollution and its associated health impacts while an example of benefits would be the impetus provided to local economies by a project.

- **Valuation of inputs and outputs.** Economic valuation uses the social value, represented by opportunity cost, rather than financial prices. For example, the social value of a fuel is its worth in an alternative use—for a fuel that could otherwise be exported then the social value is the foregone export revenue or its world market price. Where an input or output does not have a market price, such as health damages, proxy measures are used such as avoided costs of health care, the value of a statistical life, or evidence on willingness-to-pay to avoid the impacts of pollution.

- **Taxes and subsidies.** For economic valuation purposes, taxes and subsidies are generally ignored. These represent transfers between members of society and, therefore, do not increase or decrease aggregate social welfare. For example, a tax holiday for a renewable energy project represents a subsidy which is funded by an increase in taxes or a reduction in expenditures for other members of society. There is no net gain as a whole.[26]

---

[26] Taxes and subsidies may induce inefficiencies due to distortions in incentives and incur costs to administer. However, these impacts are usually ignored as being relatively small. There are also distributional impacts. These can be captured by weighting costs and benefits depending on which income groups are impacted or the assumption can be made that these are separately managed through taxes and subsidies to transfer costs and benefits between income groups within society.

- **Discount rate.** In any economic or financial analysis, future costs and incomes need to be adjusted to allow for the 'time value of money,' a process known as discounting. For financial analysis, the discount rate used is the cost of financing—for example, if an investment is funded by a loan then the discount rate equals the interest rate of the loan. For economic valuation, the social discount rate is used. This represents the cost to society of deferring consumption to future periods. It may be higher or lower than the financing cost depending on factors such as credit constraints (an inability to borrow to smooth consumption increases the relative cost of deferring consumption) and incomes (poorer countries generally have a higher preference for immediate consumption).

The application of these principles is illustrated in the calculations of economic value presented as follows.

2.2   Estimating the Economic Benefits of Renewable Power Generation

A key requirement is that *prices paid to renewables reflect but do not exceed their benefits to Indonesia.* Estimating these benefits requires an economic valuation at this stage in the interest of determining the social benefits of renewables. How these benefits are reflected in the financial prices paid to renewables generators and in the requirements for subsidies is covered later in this chapter and in following chapters.

The benefits (and costs) of renewables depend on the assumed technology, its scale, and its location (with different PLN grids having different fuel mixes, different demand–supply balances and profiles, and being located in areas with different income levels). The illustrative calculations presented here are for a 55 MW geothermal plant located on the Java–Bali grid. Similar calculations will need to be undertaken for each renewable technology and each major PLN grid to establish grid- and technology-specific benefits and, from this, limits on what should be paid to renewables generators.

2.2.1   Types of Economic Benefits

The major economic benefits of renewables projects can be broken down as follows:

- the avoided capital, fuel, and other operating costs (the "avoided economic costs") of the alternative supply source, adjusted for the relative reliability of the different technologies;

- the net pollution and resulting health impacts relative to the alternative supply source;

- the net local economic development benefits relative to the alternative supply source;

- the change in exposure to fuel price risks relative to the alternative supply source; and

- the net carbon emissions relative to the alternative supply source.

The estimation of each of these benefits is summarized in the next subsection as follows.

2.2.2   Defining the Alternative Supply Source

The starting point for the estimation of economic benefits is to define the alternative supply source to the renewables project. Ideally, this would be done through a comprehensive least-cost planning exercise which identifies how the technology and fuel mix changes with and without the renewables project and, from which, the change in direct costs can be estimated. Indeed, PLN and ESDM as regulator should adopt a more rigorous and thorough approach to system planning using modern system planning tools that would facilitate this sort of analysis for each of PLN's principal grids. At present, however, such analysis is not available.

Given this, it will be necessary to define the alternative technology and fuel based on an assessment of the likely operating characteristics of renewable generating technology options, and the current and planned composition of supply sources on the relevant grid. This will be a matter for expert judgement and, as with any matter relying on judgement, should be consulted on before the implementation of any new financing mechanism to build consensus around the assumptions applied.

A particularly important consideration in such calculations will be the extent to which variable and intermittent renewables (solar, wind, and run-of-river hydro) can be assumed to avoid the need to invest in new thermal capacity to ensure supply, as opposed to replacing energy generated by existing thermal capacity. This is a matter of intense debate in all countries. Three main approaches can be adopted to estimate the resulting "capacity credit":

- **Compare the loss of load expectation** (the risk that demand will not be met) with and without intermittent renewables being included in the generation expansion plan. This difference can be used to impute the change in unmet demand with intermittent renewables relative to fossil fuel generation and, therefore, its relative capacity credit. However, this is not feasible at this time in Indonesia given that it requires multiple simulations using an agreed and publicly available generation planning model and assumptions.

- **Estimate the contribution of intermittent renewables to meeting peak demand**, by comparing historical generation by renewables in peak hours with their installed capacity. This requires hourly (or more granular) data on renewables generation and demand by individual PLN grid which is, currently, not available on a public basis.

- **Approximate the expected generation of intermittent renewables on average as being equal to their capacity factor.**[27] This assumes that the contribution of renewables in peak hours is the same as their average contribution across all years.[28] This is the approach adopted here, given the limitations on available data, even though it is also the most simplistic.

In calculating the resulting capacity credit, we have also allowed for the unavailability of fossil fuel generation in peak hours, to ensure a fair comparison. For example, when comparing a wind generator with an average capacity factor of 34%, we allow for the expected unavailability of the alternative supply source, a gas-fired combined-cycle gas turbine (CCGT) whose average availability is 85% after planned maintenance and expected forced outages. The relative 'firmness' of a wind generator, therefore, is 0.4 times of that of a gas-fired CCGT (34%/85%) and its capacity credit or avoided economic fixed costs is calculated accordingly as 0.4 times the capacity and fixed operating and maintenance costs of a CCGT.

---

[27]   The capacity factor is defined as: generation [MWh / (installed capacity (MW) * 8760 (hours)]. It is distinct from the capacity credit, as used here, which represents the contribution of renewables to meeting peak demand relative to that of "firm" fossil fuel generation. In the absence of historical data, the capacity factors used in these calculations are obtained from Ea Analysis (2017).

[28]   This is likely to be a reasonable approximation for all technologies other than solar. In Indonesia, solar generation declines to zero by 19.00 while demand peaks at 20.00. Therefore, the contribution of solar generation to meeting peak demand is zero.

Table A2.1 provides initial proposals for PLN's Java–Bali grid on the assumed operating modes of different renewables technologies; the resulting assumption on which alternative technologies and fuels are displaced; and the assumed capacity credit. The assumptions on displaced technologies and fuels are drawn from the RUPTL 2019–2028. As shown in Figure A2.1, this includes large increases in coal and gas-fired generation taken to be indicative of their planned use to meet growth in baseload requirements and mid-merit to peaking requirements, respectively.[29]

Figure A2.1:   Java–Bali Generation Mix, 2019–2028

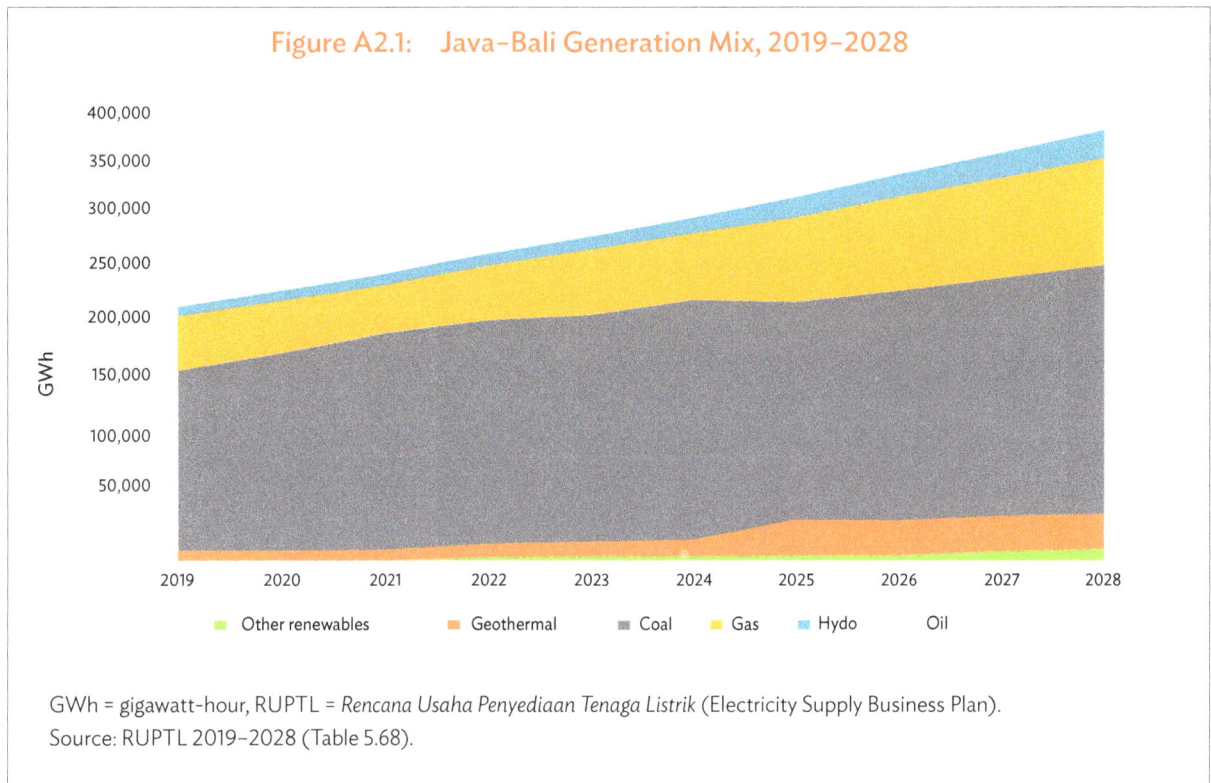

GWh = gigawatt-hour, RUPTL = *Rencana Usaha Penyediaan Tenaga Listrik* (Electricity Supply Business Plan).
Source: RUPTL 2019–2028 (Table 5.68).

It is important to note that this example takes no account of the current supply–demand balance on PLN's individual grids. Where there is excess capacity, then the benefit of adding new capacity is close to zero. Instead, the analysis presented in this chapter assumes that new capacity will be required by the time our example geothermal project comes online—which is consistent with the evidence presented elsewhere in this report of excess reserve margins declining toward and below target levels after 2023–2024 (based on capacity under construction and committed).

This example also does not consider policy or other constraints on what types of new capacity might be added but, instead, assumes that the financially least-cost option would be selected in the absence of subsidies. For example, the previous ESDM announced in 2017 that no further coal-fired power plants would be built in Java–Bali.[30] If this is considered to be official policy, then the alternative supply source for Java–Bali would inevitably change—from new coal power plants to a new gas-fired CCGT. In turn, this will change the economic benefits, the cost savings to PLN, and the maximum justified subsidy level.

---

[29] Refer to page V-45 of RUPTL 2019–2028. PLN is also planning the addition of pumped storage projects for peaking purposes. These are not considered as an alternative technology here, as it would imply replacing one set of renewables (pumped storage) with another and, therefore, would not contribute to meeting renewables expansion targets.

[30] F. Jensen. 2017. No new coal power stations in Java, Indonesia energy minister says. *Reuters*. 13 October. https://www.reuters.com/article/indonesia-power-coal/no-new-coal-power-stations-in-java-indonesia-energy-minister-says-idUSL4N1MN4ZI.

Table A2.1:   Selecting the Alternative Supply (Java–Bali Example)

	Renewables Technology	Operating Mode	Alternative Technology and Fuel	Rationale	Capacity Factor/Availability [a]		Capacity Credit [b]
					Renewables	Alternative	
1	Geothermal	All hours	Coal – USC	Operate continuously in baseload and can replace new coal capacity in this role	82.3%	79.5%	1.035
2	Solar PV	All daylight hours	Gas – CCGT	Gas CCGTs being used to meet daytime demand have a their higher fuel cost than coal, meaning they should be displaced first, and can also more easily adjust output to accommodate solar	19.4%	85.4%	0.227
3	Onshore wind	All hours with sufficient wind	Gas – CGGT / Coal – USC	Operate in both day and nighttime periods displacing fossil fuels (assumed to be gas at the margin in daytime and coal in nighttime)	34.0%	85.4% (Gas – CCGT)	0.398
4	Biomass	All hours	Coal – USC	Operate continuously in baseload and can replace new coal capacity in this role	81.5%	79.5%	1.025
5	Run-of-river hydro	All hours with sufficient water	Coal – USC	Operate in both day and nighttime periods displacing fossil fuels (assumed to be gas at the margin in daytime and coal in nighttime)	36.0% (illustrative)	79.5% (illustrative)	0.453
6	Storage hydro	At hours of highest value	Oil – GT / Gas – CCGT	Stored water is initially used to displace the alternative with the highest fuel cost (oil). If hydro output exceeds the oil-fired generation at any time, then the next-highest fuel cost generators are gas CCGTs	36.0% (illustrative)	92.2% (Oil – GT) (illustrative)	0.390

CCGT = combined-cycle gas turbine, GT = gas turbine, PV = photovoltaic, USC = ultra-supercritical

[a] For intermittent renewables, the average capacity factor is shown. For other technologies, the availability after allowing for planned and expected forced outages is shown.
[b] The capacity credit is calculated as the capacity factor and/or availability of the renewables technology divided by that of the alternative supply source.

Source: ADB, using data from Ea Analysis (2017).

2.2.3   Avoided Economic Costs

Avoided economic costs are a function of the assumptions made on the alternative technology and fuel to the renewables project being considered. For our example 55 MW geothermal project on the Java–Bali grid, we assume the alternative is a new ultra-supercritical coal-fired power plant operating in baseload mode. The resulting calculation of the avoided direct costs over the lifetime of the power plant is summarized in Table B2. As described earlier, this calculation uses economic values, with key assumptions being that

- the economic cost of coal is equal to the 12-month average world market price of $86.9/ton;[31] and

- the social discount rate used for present value estimates is 9%, generally applied by Asian Development Bank (ADB) for energy sector projects, in the absence of official Indonesian values.[32]

Table A2.2 also shows the availability adjustment factor and resulting avoided cost. This exceeds the direct cost, due to the assumed availability factor of geothermal power plants being slightly higher than that of ultra-supercritical power plants (Table A2.1).

Table A2.2:   Calculation of Avoided Direct Costs (Geothermal Against Coal)

Cost Item	Unit	Alternative Technology: Coal Ultra-Supercritical (thermal efficiency = 42%)		
		Direct Costs of Coal Power Plant	Availability Adjustment Factor	Avoided Direct Costs
CAPEX	USC/kWh	2.43	1.035	2.51
Fixed O&M	USC/kWh	0.81	1.035	0.84
Variable O&M	USC/kWh	0.01	--	0.01
Fuel	USC/kWh	2.97	--	2.97
Total	USC/kWh	6.22		6.33

CAPEX = capital expenditure, O&M = operation and maintenance, USC/kWh = US dollar cents per kilowatt-hour.
Source: ADB calculations.

---

[31]   An alternative would be to use forecast prices. However, this raises questions as to the veracity and reliability of such forecasts where historical prices are less subject to dispute. For a mechanism intended to determine eligibility for subsidies, it is obviously desirable to minimize potential areas of disagreement.

[32]   Some countries determine social discount rates to be used in public sector analysis. For example, in the United Kingdom, this and other guidance are contained in "The Green Book" published by the Treasury. This current advises the use of a social discount rate of 3.5% in real terms, which compares with a current long-term real yield on government bonds of 0.5%. Existing government documents on low-carbon paths for Indonesia have been reviewed but unable to identify any social discount rate applied in cost–benefit analysis of these paths. HM Treasury (UK). 2018. *The Green Book: Central Government Guidance on Appraisal and Evaluation.* https://assets.publishing.service.gov.uk/government/uploads/system/uploads/attachment_data/file/685903/The_Green_Book.pdf. The Government of Indonesia documents reviewed: (i) Ministry of Finance (Indonesia). 2009. *Green Paper: Economic and Fiscal Policy Strategies for Climate Change Mitigation in Indonesia* and (ii) Badan Perencanaan Pembangunan Nasional (BAPPENAS, National Development Planning Agency). 2019. *Low Carbon Development: A Paradigm Shift Towards a Green Economy in Indonesia.*

2.2.4   Pollution and Health Benefits

Replacing coal-fired generation with geothermal generation will have corresponding health and environmental benefits through the resulting reduction in emissions of harmful pollutants. The following levels of emissions from coal-fired power plants are assumed, corresponding to currently permissible maximums:[33]

- Particulate Matter (PM)$_{2.5}$: 150 milligrams per cubic meter (mg/Nm^3) (for an ultra-supercritical coal-fired power plant, this is equivalent to 0.752 grams per kilowatt hour (g/kWh)

- Nitrogen oxide (NO$_x$): 263 g/GJ fuel (equivalent to 2.254 g/kWh)

- Sulfur dioxide (SO$_2$): 350 g/GJ fuel (sulphur) with 73% abatement (equivalent to 1.620 g/kWh with abatement)

Damage costs associated with emissions of these pollutants are obtained from IMF estimates for 2014,[34] updated to current Indonesian gross domestic product values. These are:

- PM$_{2.5}$: 6,914 USD per tonne ($/t) (for an ultra-supercritical coal-fired power plant, this is equivalent to 0.52 US dollar cents/kilowatt-hour [USC/kWh])

- NO$_x$: 3,057 $/t (equivalent to 0.69 USC/kWh)

- SO$_2$: 5,664 $/t (equivalent to 0.92 USC/kWh)

No emissions of pollutants are assumed from geothermal power plants.

2.2.5   Local Economic Development

The relative contributions of geothermal and the alternative coal-fired power plants to the economic development of their host communities are estimated as a function of capital expenditures, of which 25% is assumed to be spent locally reflecting civil works, and nonfuel operating expenditures, of which 75% is assumed to be spent locally, reflecting personnel costs. As geothermal plants do not have fuel costs, a higher proportion of their total costs is consequently assumed to be spent locally (32.6% of all costs compared to 19.7% of the costs of an ultra-supercritical coal-fired power plant). These local expenditures are assumed to trigger additional local economic development (e.g., in associated housing services) equal to 75% of the expenditures, a multiplier of 0.75.[35] For example, for every $1,000 spent on a geothermal power plant, $326 is assumed to be spent locally and triggers additional local economic development equivalent to $245.

The value of local economic development mobilized by power plants is estimated as equivalent to 2.42 USC/kWh for a geothermal power plant compared to 0.92 USC/kWh for an ultra-supercritical coal-fired power plant. The difference of 1.50 USC/kWh represents the net benefit of a geothermal power plant arising from this source.

---

[33]   Minister of Environment Regulation 21/2008.

[34]   Parry I et al. 2014. *Getting Prices Right: From Principle to Practice.* IMF (http://www.greenfiscalpolicy.org/wp-content/uploads/2014/11/Getting-Energy-Prices-Right-Full-Publication.pdf; the accompanying database can be downloaded at https://www.imf.org/external/np/fad/environ/data/data.xlsx]. Damage costs were updated to 2020 values using Organisation for Economic Co-operation and Development (OECD) data.

[35]   This is the value used in previous economic analysis conducted by ADB and the World Bank to support proposals on reforms to geothermal policies and regulations. ADB and World Bank. 2015. *Unlocking Indonesia's Geothermal Potential.* Manila. https://www.adb.org/sites/default/files/publication/157824/unlocking-indonesias-geothermal-potential.pdf.

2.2.6   Reduced Exposure to Fuel Price Risk

Development of renewables provides a means to reduce Indonesia's exposure to fossil fuel price risks, which imposes costs on Indonesia due to measures taken to stabilize electricity prices and offset the impacts of higher fuel prices. This benefit is estimated at 0.26 USC/kWh, which is the average difference between an imputed forecast (based on trailing average prices) and actual spot prices over the last 10 years.[36] This can be interpreted as the cost of managing unexpected deviations of actual fuel prices from forecast levels. These have real costs as, for example, they may lead to government imposing caps on fuel prices which, in turn, result in lost revenue for the country.

2.2.7   Carbon Emissions

The final benefit considered is that of reductions in carbon emissions. The assumption applied here is that Indonesia has made internationally public commitments to reduce emissions, in the form of its Nationally Determined Contribution submitted under the Paris Agreement. Therefore, Indonesia faces a cost (even if only reputational, at present) if it does not meet its targets for reductions.

- At present, Indonesia does not itself calculate a social cost of carbon for use in economic analysis of the type presented here. Therefore, for the purposes of estimating the value of reductions in emissions resulting from generation from our example geothermal power plant as opposed to the alternative coal-fired power plant, we use the ADB's default social cost.[37] As of 2020, this is \$39.3/ton of carbon dioxide equivalent (tCO_{2e})[38] which, given estimated carbon dioxide (CO_2) emissions from an ultra-supercritical coal-fired power plant of 0.794, gives a benefit from reduced emissions of 3.12 USC/kWh in that year. This increases over time with a rising social cost of carbon, with the levelized benefit over the lifetime of the power plant being 4.05 USC/kWh.

2.2.8   Total Economic Benefits

The estimated total economic benefits of a geothermal power plant replacing an ultra-supercritical coal-fired power plant on the Java–Bali grid are estimated at 14.27 USC/kWh or Rp2,018/kWh, broken down as shown in Table A2.3. These represent the benefits from this specific case. As noted earlier, the benefits will differ by renewables technology, and by grid, which will define the alternative technology and fuel type and, therefore, the benefits realized.

It is worth noting that:

- Less than half of the economic benefits arise from the avoided economic costs of the alternative technology. By implication, this implies a large mismatch between the social benefit to Indonesia of renewables and the cost savings to PLN—a point explored further later in this report in discussing subsidy needs.

---

[36]   The calculation uses World Bank commodity price data. The methodology follows that in ADB and World Bank (2015).

[37]   This is \$36.3/tCO_{2e} in 2016, increasing in real terms by 2% annually. See para. 163 in: ADB. 2017. Guidelines for the Economic Analysis of Project. https://www.adb.org/sites/default/files/publication/157824/unlocking-indonesias-geothermal-potential.pdf.

[38]   BAPPENAS (2019) includes a link to a source for country-specific estimates of the social cost of carbon, which values this at \$10.9/ tCO_{2e} for Indonesia (https://country-level-scc.github.io/explorer/). However, it does not appear to endorse the use of this value for cost–benefit analysis for Indonesia.

- Reductions in carbon emissions represent over one-quarter of the total economic benefits. If these are set at zero, as might be argued to be appropriate for a developing country whose commitments to reduce emissions are not legally binding on it, then the economic benefit reduces to 10.22 USC/kWh.

Table A2.3:   Estimated Economic Benefits
(Geothermal on Java–Bali Grid)

Benefit	Value		Share of Total Benefit
	USC/kWh	Rp/kWh	
Avoided direct costs	6.33	895	44%
Pollution and health impacts	2.13	301	15%
Local economic development	1.50	212	11%
Reduced fossil fuel price risk	0.26	37	2%
Carbon emissions	4.05	573	28%
Total	14.27	2,018	100%

kWh = kilowatt-hour, Rp = Indonesian rupiah, USC = United States dollar cents.

Source: ADB calculations.

2.3   Estimated Financial Costs

The preceding calculations estimate the economic benefits to Indonesia of developing renewables (in this case, a geothermal power plant) relative to the alternative fossil fuel generator. However, in order to discuss potential subsidy needs, two further pieces of information are required:

- **What are the financial costs of the geothermal project?** This represents what the developer needs to recover. If these exceed the economic benefit, then the conclusion must be that the costs to Indonesia outweigh the benefits.

- **What the financial costs saved by PLN?** This represents the price for renewables purchases that PLN can absorb without its costs and retail tariffs increasing. If this is more than the financial costs of the geothermal project then there is no issue. However, if these costs are less than the price required for the geothermal project to recover its costs then there is a need for either retail tariffs to increase or subsidies to cover the gap.

An initial estimation of these two sets of financial costs and comparison with the estimated economic benefits to Indonesia, for our example geothermal project replacing coal-fired generation on the Java-Bali grid, is presented as follows.

2.3.1 Geothermal Financial Costs

Our estimate of the financial costs of a new geothermal power plant is based on calculations prepared for PT Sarana Multi Infrastruktur (SMI),[39] assuming private sector financing. Key assumptions are summarized in Table A2.4.

Table A2.4:   Summary of Geothermal Cost Estimates

Item	Estimate	Item	Estimate
Capacity	55 MW	Life	30 years
Capacity factor	82.3%	Financing cost (nominal)	12.5% (nominal) 10.3% (real)
Investment cost ($/kw)	5,330 $/kW (includes exploration and delineation)	Make-up and replacement wells	$25.5 million every 10 years

kW = kilowatt, MW = megawatt.

Source: ADB. Capacity factors allow for scheduled and forced outages.

The estimated financial cost of a new geothermal power project, located in a moderately accessible location in Java, is 10.97 USC/kWh in real terms (excluding inflation) or 14.63 USC/kWh including inflation. This compares to estimated economic benefits of 14.27 USC/kWh, in real terms.

2.3.2 Cost Savings

PLN's cost savings represent the avoided financial costs of the alternative supply source—in our example, that of a new coal-fired power plant commissioning post-2023 when excess reserve margins start to disappear without new investment. The calculation of these avoided costs is the same as that for the avoided direct costs incorporated into the economic benefit calculations, with two differences:

- **Fuel costs.** The calculation of financial cost savings assumes a continuation of the current policy capping the price of domestic coal supplied to PLN at $70/t. This is below the 12-month average price on international markets, meaning that PLN's financial cost of fuel is less than the economic cost to Indonesia.

- **Cost of financing.** The calculation of economic costs uses a social discount rate of 9% for the purposes of estimating present values. The financial costs calculation, as with the estimation of the costs of a geothermal project presented above, uses the estimated actual cost of financing of PLN. The reliance of PLN on government support means that its perceived risk and cost of financing is similar to that of the Indonesian government.[40] Therefore, PLN's financing cost is estimated as the return on long-run

---

[39]   GT Management. 2019. *Cost of Production from Geothermal Power Projects in Indonesia*. PT SMI. Funded by New Zealand Foreign Affairs and Trade Aid Programme.

[40]   For example, in October 2018, Moody's affirmed PLN's credit rating as Baa2, equal to Indonesia's sovereign credit rating. In doing so, Moody's noted that: "...*the rating reflects the strategic importance of PLN, and our expectation of a very high likelihood of government support in a distressed situation. Such expectation of support considers the 100% government ownership in PLN, plus the strategically important role that PLN plays in Indonesia's critical power sector.*" Moody's. 2018. *Moody's affirms PLN's Baa2 ratings: Outlook remains stable.* https://www.moodys.com/research/ Moodys-affirms-PLNs-Baa2-ratings-Outlook-remains-stable--PR_390040.

government bonds, currently at 7.1%,[41] plus a margin of 100 basis points. This gives a cost of financing of 8.2% in nominal terms[42] or 6.0% in real terms.[43]

With these adjustments, the estimated financial costs of a new coal-fired power plant located on the Java–Bali grid, and financed by PLN, are 4.95 USC/kWh in real terms or 6.74 USC/kWh in nominal terms. This compares to the current average generation production cost (BPP) for Java–Bali of approximately 7.0 USC/kWh (a nominal value).

That the estimated financial cost savings to PLN are below the generation BPP for the Java–Bali grid can be explained by a number of factors:

- The generation BPP represents the average cost for PLN. This includes higher-cost generation which is not substituted by geothermal (e.g., oil-fired generation only used in peak hours). By comparison, the financial cost savings represent the marginal cost of meeting baseload demand—the mode in which geothermal operates.

- In circumstances such as those on the Java–Bali grid at present, where there is excess capacity, then average cost is likely to be above marginal cost due to the inclusion of the costs of paying for capacity that is not needed to meet demand.

- The calculation of financial cost savings assumes that PLN is financing generation investment whereas the BPP includes a mix of PLN-financed generation and privately financed generation (IPPs). For comparison, if applying the same cost of financing as for new geothermal power plants, then the financial cost savings to PLN of replacing new coal power plants rises to 7.83 USC/kWh in nominal terms, or above BPP.

What this comparison does highlight is two important issues for the calculation of permissible subsidies:

- The generation BPP is not a good measure of the financial cost savings to PLN, which are driven by the costs of the alternative future supply source and not legacy investments. In some grids this will be higher and in some lower than the generation BPP.

- There is a need for a clear and consistent set of principles to be used in calculating PLN's financial cost savings to ensure like-for-like comparisons with the costs of renewable energy technologies and with economic benefits to Indonesia. These principles will need to form part of any implementing regulation.

[41] As of 25 October 2019, the yield on Indonesia's 10-year bond was 7.09% (Wall Street Journal).
[42] The Fisher formula is applied: Return = (1 + 7.1%) * (1 + 1.0%) - 1
[43] The year-on-year change in the Indonesian Consumer Price Index (CPI) for September 2019 was 3.39% (Bank Indonesia).

2.4   Economic Benefits and Financial Costs Compared

Figure A2.2 compares the economic benefits of our example geothermal project with its financial costs and with the financial costs saved by PLN. As can be seen, the cost of the project is substantially below its economic benefits, implying a large gain to Indonesia. However, its financial cost is also far above the cost savings to PLN, implying that proceeding with the project requires either retail tariff increases to cover the difference or a subsidy payment to close the gap.

Figure A2.2:   Economic Benefits and Financial Costs
(Java–Bali Geothermal Replacing Coal)

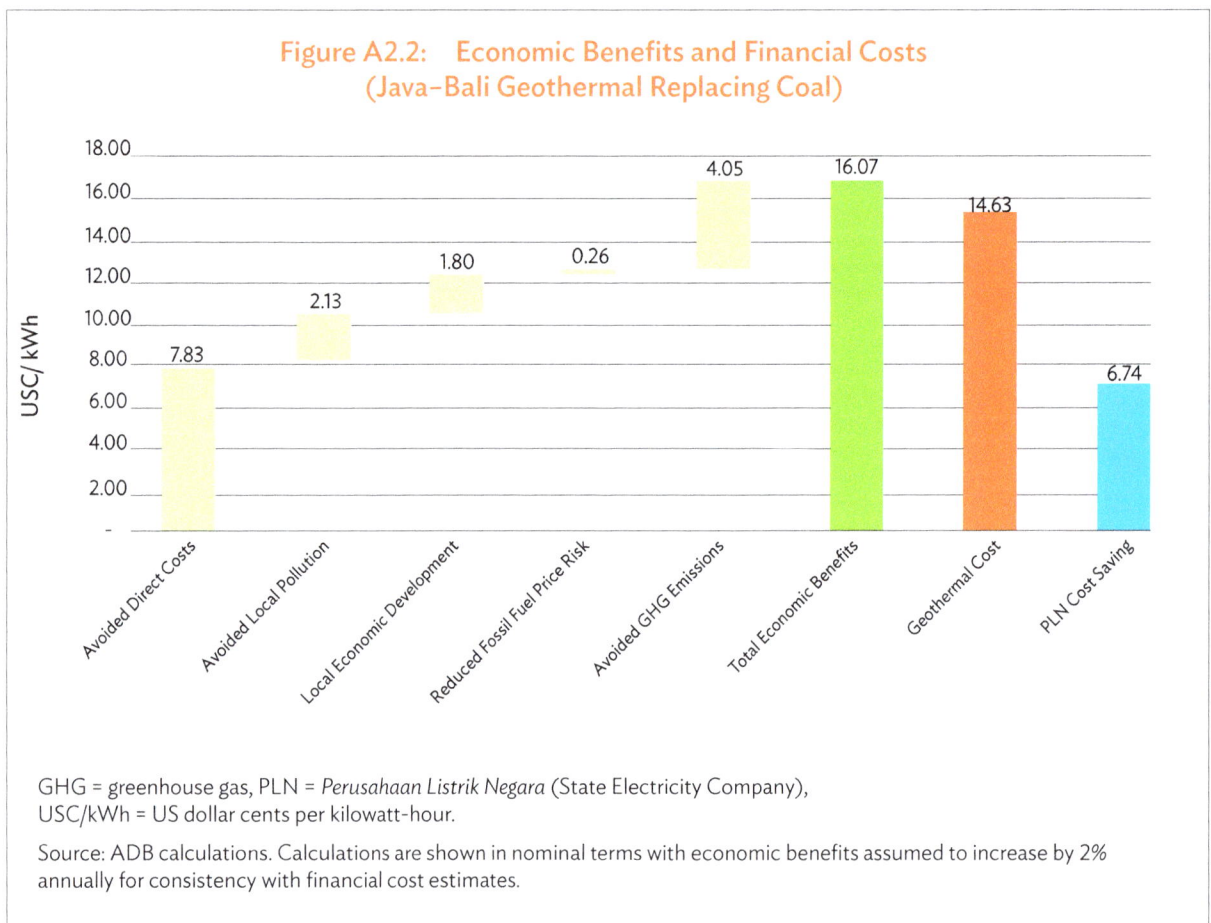

GHG = greenhouse gas, PLN = *Perusahaan Listrik Negara* (State Electricity Company),
USC/kWh = US dollar cents per kilowatt-hour.

Source: ADB calculations. Calculations are shown in nominal terms with economic benefits assumed to increase by 2% annually for consistency with financial cost estimates.

2.5   Subsidies and Geothermal Development

This concluding section considers the scale of the potential subsidy requirements for geothermal energy under the proposed mechanism and the expected quantities delivered. As described in Section 3.10, two scenarios are used. The first represents the planned additions under the RUPTL 2019–2028 and the second represents an "adjusted" scenario. This removes capacity from projects not yet identified given the required lead time and reduces capacity estimates to match more realistic estimates of resource potential.

The two supply curves are illustrated (Figure A2.3) compared to the estimated avoided cost to PLN and the economic value as at 2025 (the date used for establishing target renewable energy shares).[44] As all identified projects are located on the Java–Bali or Sumatera grids, where the alternative baseload capacity is assumed to be modern coal power plants, the avoided cost and economic values are the same for all projects.

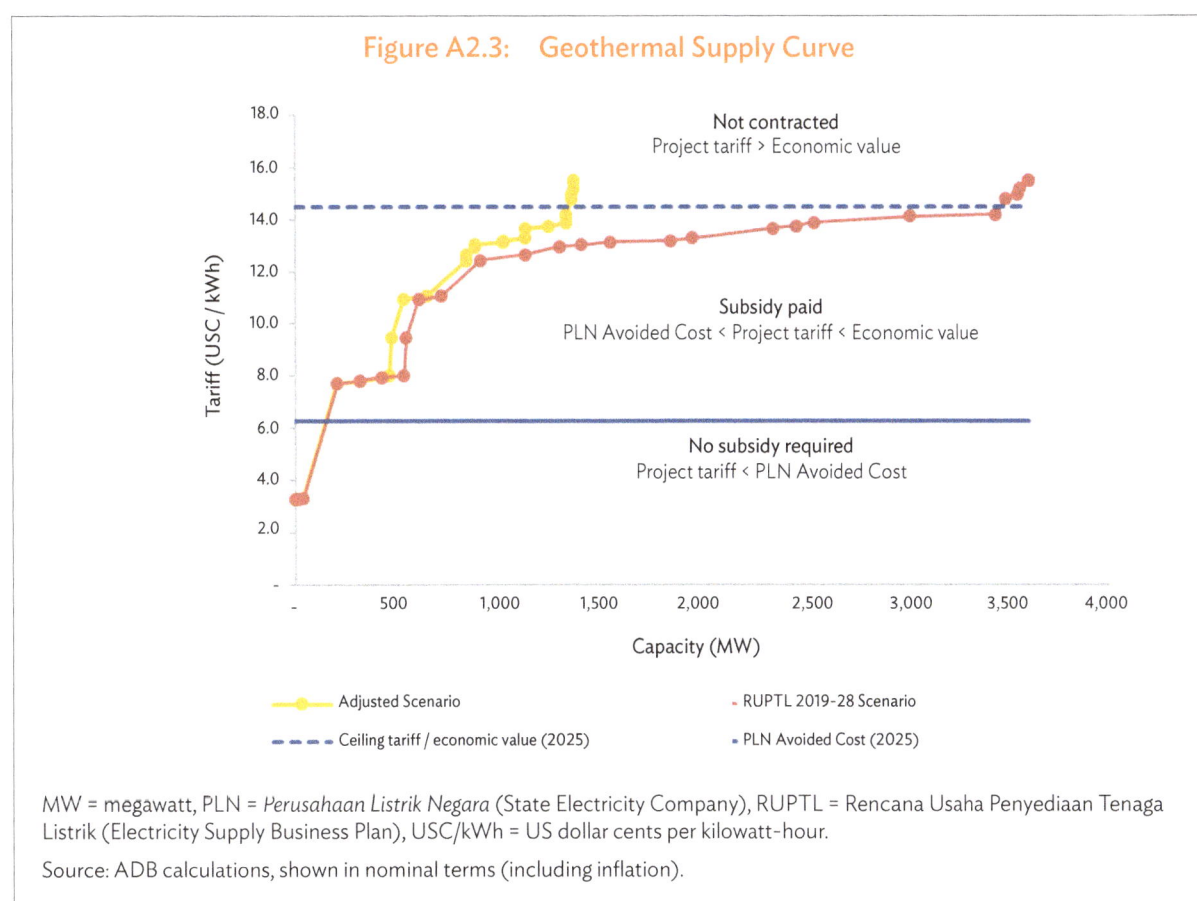

Figure A2.3:   Geothermal Supply Curve

MW = megawatt, PLN = *Perusahaan Listrik Negara* (State Electricity Company), RUPTL = Rencana Usaha Penyediaan Tenaga Listrik (Electricity Supply Business Plan), USC/kWh = US dollar cents per kilowatt-hour.

Source: ADB calculations, shown in nominal terms (including inflation).

A comparison of the results is provided in Table A2.5, immediately following Figure A2.3. Under the RUPTL scenario, annual subsidies of Rp18.0 trillion mobilize an additional 3,318 MW of geothermal capacity, generating 26,784 GWh annually. Under the adjusted subsidies, annual subsidies fall to Rp6.4 trillion but only 1,200 MW of additional capacity is mobilized, generating 9,687 GWh. The average subsidy per unit of geothermal energy delivered is similar across the two scenarios.

---

44   The avoided cost and economic value estimates change annually with forecast fuel prices, technology cost changes and inflation. The project tariff is compared with the avoided cost and economic value in the expected year of commercial operations.

Table A2.5:   Additional Geothermal Capacity Deployment and Subsidies

	Number	Deployment		Subsidies	
		Capacity (MW)	Generation (GWh)	Annual Total (Rp trillion)	Average (USC/kWh)
RUPTL 2019–2028 Scenario					
Projects without subsidy	2	40	323		
Projects with subsidy	18	3,318	26,784	18.04	4.77
Projects not contracted	5	246	1,986		
Adjusted Scenario					
Projects without subsidy	2	40	323		
Projects with subsidy	12	1,200	9,687	6.38	4.66
Projects not contracted	5	126	1,017		

GWh = gigawatt-hour, MW = megawatt, Rp = Indonesian rupiah, RUPTL = *Rencana Usaha Penyediaan Tenaga Listrik* (Electricity Supply Business Plan), USC/kWh = US dollar cents per kilowatt-hour.

Source: ADB calculations.